Intestinal Water and Electrolyte Transport in Health and Disease

Colloquium Series on Integrated Systems Physiology: from Molecule to Function to Disease

Physiology is a scientific discipline devoted to understanding the functions of the body. It addresses function at multiple levels, including molecular, cellular, organ, and system. An appreciation of the processes that occur at each level is necessary to understand function in health and the dysfunction associated with disease. Homeostasis and integration are fundamental principles of physiology that account for the relative constancy of organ processes and bodily function even in the face of substantial environmental changes. This constancy results from integrative, cooperative interactions of chemical and electrical signaling processes within and between cells, organs, and systems. This eBook series on the broad field of physiology covers the major organ systems from an integrative perspective that addresses the molecular and cellular processes that contribute to homeostasis. Material on pathophysiology is also included throughout the eBooks. The state-of the-art treatises were produced by leading experts in the field of physiology. Each eBook includes stand-alone information and is intended to be of value to students, scientists, and clinicians in the biomedical sciences. Since physiological concepts are an ever-changing work-in-progress, each contributor will have the opportunity to make periodic updates of the covered material.

Published titles

(for future titles please see the Web site, www.morganclaypool.com/page/lifesci)

Intestinal Water and Electrolyte Transport in Health and Disease
Mrinalini C. Rao, Jayashree Sarathy (nee Venkatasubramanian), and Mei Ao
www.morganclaypool.com

ISBN: 9781615041565 paperback

ISBN: 9781615041572 ebook

DOI: 10.4199/C00049ED1V01Y201112ISP031

A Publication in the

COLLOQUIUM SERIES ON INTEGRATED SYSTEMS PHYSIOLOGY: FROM MOLECULE TO FUNCTION TO DISEASE

Lecture #31

Series Editors: D. Neil Granger, LSU Health Sciences Center, and Joey P. Granger, University of Mississippi Medical Center

Series ISSN

ISSN 2154-560X print

ISSN 2154-5626 electronic

Intestinal Water and Electrolyte Transport in Health and Disease

Mrinalini C. Rao
University of Illinois

Jayashree Sarathy (nee Venkatasubramanian)
University of Illinois
Benedictine University

Mei Ao
University of Illinois

COLLOQUIUM SERIES ON INTEGRATED SYSTEMS PHYSIOLOGY: FROM MOLECULE TO FUNCTION TO DISEASE #31

ABSTRACT

The unique architecture and physiology of the mammalian intestine, together with a tightly coordinated regulatory system, allows for the handling and absorption of as much as 9 L of fluid a day with 98% or greater efficiency. Advances in the past 40 years have made inroads into revealing the intricacies and interplay of numerous ion transporters and their modulators that are responsible for intestinal electrolyte and water transport. Studies of two devastating diseases, the virulent infectious disease cholera and the autosomal recessive disease cystic fibrosis, were largely responsible for this information explosion. These advances have been critical in the development of new therapeutic strategies to combat life-threatening diseases of varying etiologies ranging from enteric infections to cystic fibrosis and inflammatory bowel diseases. Yet, the story is far from complete, and progress needs to continue on translating information gained from reductionistic cell and tissue culture models, in vivo models, and ultimately human studies and on improving therapeutic approaches. This book reviews the current status of our knowledge of fluid transport across the intestine, including the complexities of transcellular and paracellular ion transport down the length of the intestine and how aberrations of normal physiological processes lead to disease.

KEYWORDS

intestinal architecture, epithelial cell, transporters, pumps and channels, transepithelial transport, electrolyte and water movement, anion and cation transport, regulation of intestinal transport, intestinal disorders

Contents

CHAPTER 1

Overview

The mammalian intestine is a complex, intricately regulated organ system that is structurally and functionally geared to efficiently digest and absorb nutrients, to ensure nonabsorbable substances are effectively removed from the body, and to provide a protective role against noxious external stimuli. A critical role of the healthy adult mammalian intestine is that it handles as much as 9 L of fluid a day, including dietary intake (~2 L) and secretions emanating from salivary glands (~1.5 L), gastric mucosa (~2.5 L), pancreas (~1.5 L), bile and the intestine (~1.5 L) (Figure 1). It does so with great efficiency such that only about 100–200 mL of this fluid is lost in the stool daily. An oft-posed question is: Does the intestine itself need to secrete 1.5 L of additional fluids? This secretion is necessary to help in the proper mixing of chyme with hydrolytic enzymes, to allow for maximal and optimal exposure of the resulting nutrient substrates to the absorptive surface, for easy transit down the cephalocaudal axis with additional absorption in the colon and for the formation and passage of stool. By their lubricating action, intestinal secretions may help prevent the abrasions and epithelial damage caused by mechanical stress.

To cope with its various functional demands, including electrolyte and water transport, the intestinal architecture changes down the length of the cephalocaudal axis as well as exhibits distinctive features along the crypt–villus or crypt–surface axes of the small and large intestine, respectively. Although net fluid movement can be influenced by motility of the muscle layers, the epithelial layers lining the small intestinal and colonic lumen are the sites of electrolyte and water transport. The polarized enterocytes and colonocytes possess an assortment of transporters that exhibit complex interactions with the cytoskeleton and intracellular modulators to respond to changes in the external milieu and render vectorial transport. These epithelial ion transport processes are also important for maintaining cell membrane potential, ionic composition, volume, and pH and play critical roles in cell growth, differentiation, and even cell death. Epithelial function is critically dependent on cell polarity, and intercellular junctional complexes help maintain this polarity and influence transit through the paracellular route.

The cues for orchestrating these processes arise from a vast array of regulators. Hormones, neurotransmitters, immunomodulators, and paracrine and autocrine mediators contributed by systemic

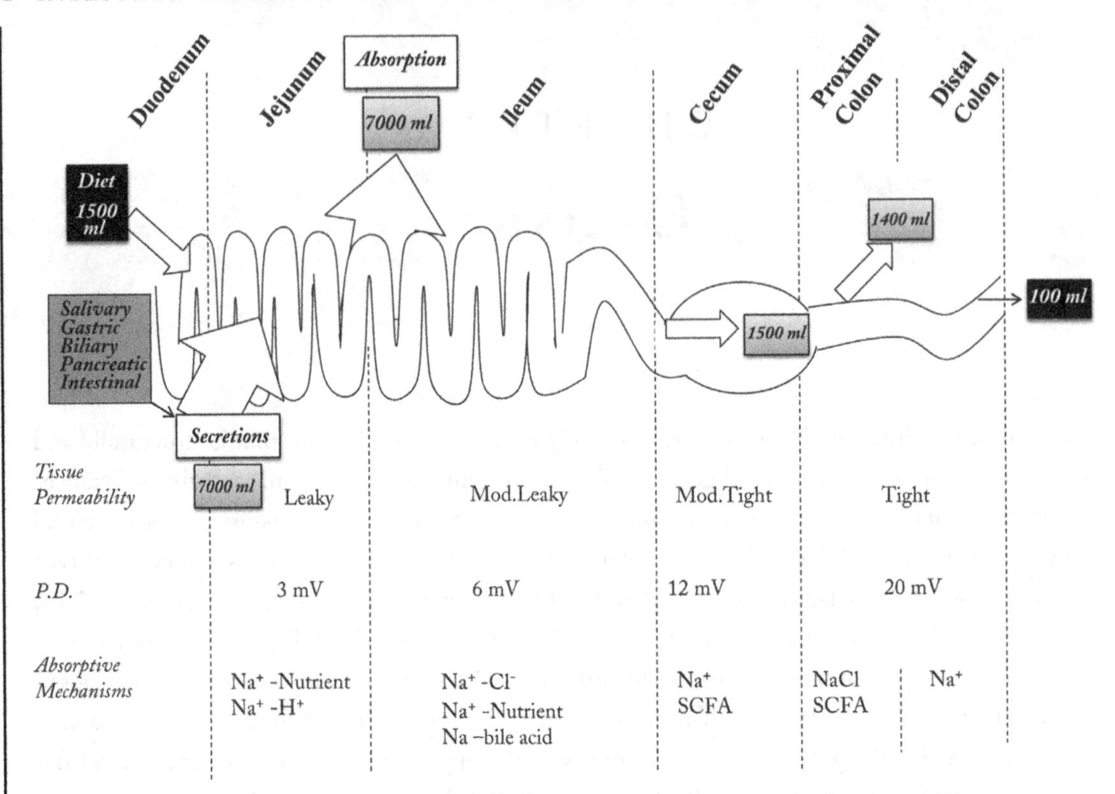

FIGURE 1: Overview of intestinal fluid balance. The average fluid flow into the intestine is 8 to 9 L. The amount is made up of salivary, gastric, biliary, pancreatic, and intestinal secretions, and dietary intake. The small intestine absorbs most of the fluid and about 1.5 L of fluid flows to the colon, where fluid is further absorbed, leaving only 100 to 200 mL of fecal water daily. SCFA, short-chain fatty acids. Modified from Ref. 114 (Sleisenger and Fordtran's Gastrointestinal and Liver Disease, Chapter 99; permission obtained).

and local endocrine cells, neural circuitry, and lymphoid tissues, as well as the commensal microbiota residing in the intestinal lumen, contribute to gut function. It is therefore not surprising that there exists an intricate interplay not only between the transport processes but also an immense amount of cross-talk in the signaling cascades that regulate these processes; furthermore, there are some inherent and necessary overlaps of function and redundancies in regulation. It is also not difficult to conceive that such a multifaceted system is susceptible to aberrations and perturbations and can lead to disease. It is indeed the intense study, over the past four decades, of the pathophysiology underlying two devastating diseases, cholera and cystic fibrosis (CF), that have led to our current understanding of intestinal electrolyte and water transport at the cellular and molecular levels.

Cholera, a disease caused by a virulent bacterium causes copious amounts of fluid secretion due to uncontrolled production of cyclic adenosine monophosphate (cAMP), whereas CF, an autosomal recessive disease, is at the other end of the spectrum leading to insufficient or minimal secretion due to an aberrant Cl^- channel. More recently, scrutiny of the mutifactorial components underlying the broad etiologies of inflammatory bowel diseases (IBDs) is unraveling new and important information about the complexities of both cellular and paracellular transport. A variety of approaches ranging from genetically engineered animal models, to in vitro cellular and biochemical strategies, to sophisticated cell and whole animal imaging, and translation into human studies have provided critical tools in developing new therapeutic strategies for treating a wide variety of life-threatening intestinal disorders manifesting as diarrhea or constipation. A medical marvel of the twentieth century with a wide global impact, which is based on an understanding of fundamental physiological process, is oral rehydration therapy for the treatment of diarrheal diseases. In the case of CF, rigorous pancreatic enzyme supplement regimens and use of inhibitors and antibiotics to relieve respiratory distress have vastly improved the life expectancy over the past four decades. Newer methodologies are using combinatorial chemistry to design highly specialized drugs that can focus on a specific intestinal disorder while attenuating untoward side effects.

In this book, we provide the reader with an overview of our current understanding of the cellular and molecular basis of electrolyte and water transport along the length of the intestine and how these are affected in diseases states. Like many areas of human physiology, there has been an information explosion in the field as the molecular entities (e.g., multiple isoforms and related anion exchangers [AEs]) underlying well-documented function (e.g., anion exchange) are being discovered. The better we understand the nuances of normal intestinal function, the better will be our ability to develop precise mechanisms for therapy and disease management. For the sake of convenience, most of the literature cited is in the form of topical reviews.

• • • •

CHAPTER 2

Epithelial Cell and Tissue Architecture

2.1 POLARIZED EPITHELIAL CELL

The hallmark of intestinal epithelial cell function, vectorial movement of fluid and electrolytes, is dependent on the cells possessing their polarity (Figure 2). The intestinal epithelial tissue is a columnar epithelium (with the exception of the rectum, where it is a stratified squamous epithelium), and its polarity is defined by cells held in close apposition by tight junctions (TJs) and each cell possessing two discrete plasma membrane domains—the apical membrane facing the lumen and the basolateral membrane facing the serosal or blood side. Each of these membranes has distinct protein and lipid compositions and therefore distinct biophysical properties leading to different functions. Vectorial transport across all epithelia is energized by the Na^+/K^+-ATPase, which utilizes ATP to maintain a negative intracellular voltage and a low intracellular Na^+ ($[Na^+]_i$) and is inhibited by the cardiac glycoside, ouabain. With the exception of the choroid plexus, where it is located on the apical membrane, the Na^+/K^+-ATPase is located on the basolateral membrane of all epithelia.

The characteristic features of a specific epithelial segment are further defined by the nature of the TJs and the variety of different proteins inserted into their apical or basolateral membrane. The TJs are dynamic structures serving both as a fence and as a gate-a molecular fence in that they restrict the diffusion of proteins and lipids between the apical and basolateral membranes and a gate in that they restrict the movement of electrolytes between the luminal and the serosal compartments through the extracellular space. Early studies on epithelial polarity demonstrated that when TJs are disrupted in ex vivo preparations, there is a time-dependent and discernable diffusion and intermingling of proteins through the fluid phase of the apical and basolateral membranes. The net result is a loss of epithelial cell polarity. For example, Na^+/K^+-ATPase has been shown to redistribute to the apical membrane in renal allografts during postischemic injury [60]. However, other evidence suggests that the effects of TJ disruption are much more nuanced and certainly are cell- and tissue-specific. For example, Mandel et al. [72] demonstrated that while ATP depletion disrupts the gate function of TJs, causing transepithelial resistance (TER) to drop precipitously, the fence maintains the asymmetry of apical and basolateral membrane lipid, and the protein composition is not lost

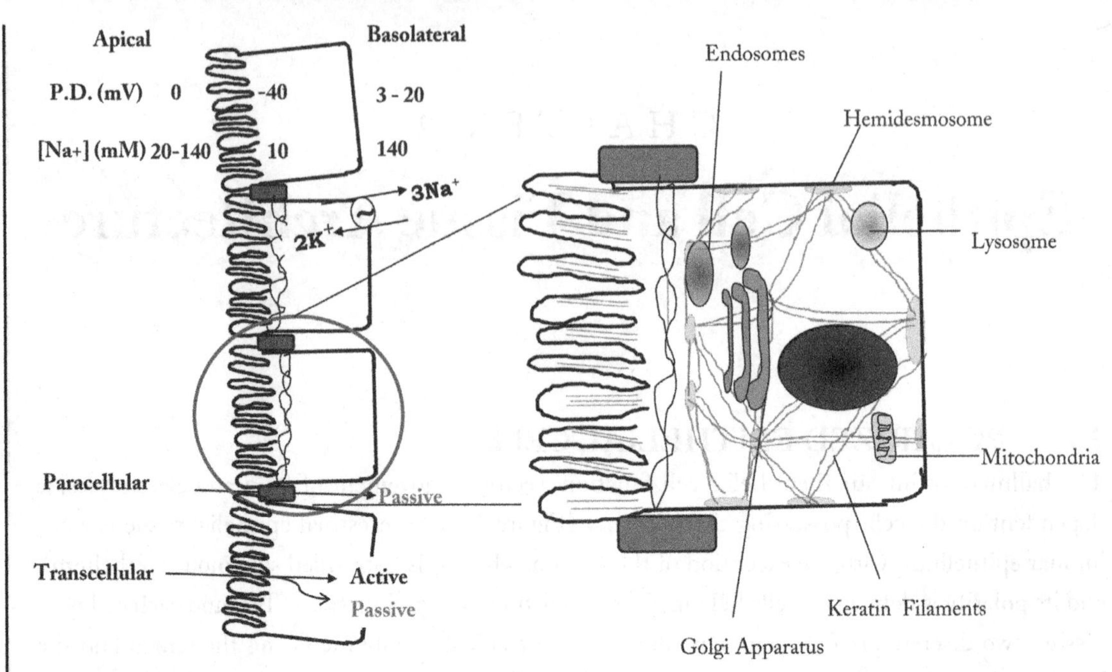

FIGURE 2: Intestinal epithelial cell. Intestinal epithelial cells are well-designed for vectorial transport: (1) the cell membrane is divided into distinct apical and basolateral zones by the tight junctions with an asymmetric distribution of transporters; (2) there is an electrochemical profile across the epithelium that permits "downhill" entry of sodium into the cell from either side of the cell; (3) the Na^+/K^+ ATPase (sodium pump) on the basolateral membrane is essential for maintaining the electrochemical profile; and (4) water and solutes can cross the epithelium either around the cell (paracellular) or through the cell (transcellular). P.D., potential difference. Modified from Ref. 114 (Sleisenger and Fordtran's Gastrointestinal and Liver Disease, Chapter 99; permission obtained).

over the same period. Protein–protein interactions, including those with cytoskeletal scaffolding proteins, contribute to this maintenance of the fence function.

Advances in sophisticated microscopy and live cell imaging techniques have led to our current understanding of how a newly synthesized epithelial membrane protein is sorted to the appropriate target membrane. Many proteins possess specific sorting signals, and in combination with intracellular trafficking processes involving vesicular transport and cytoskeletal elements, there are highly regulated mechanisms to assure that the protein is inserted in the correct membrane. Thus, proteins destined for the basolateral membrane contain specific amino acid sequences in their cytoplasmic tails that are identified as membrane-sorting signals. In contrast, membrane proteins destined for the apical membrane, such as the well-characterized apical membrane marker alkaline phosphatase, possess a glycosyl phosphatidyl inositol anchor, which allows their association with lipid rafts and insertion into the target membrane [2]. There are some proteins that can be held at the basal pole of the cell by cytoskeletal proteins such as ankyrin before being inserted into their target apical or basolateral membrane. Yet other proteins are randomly inserted into both apical and basolateral membranes [82]. Since proper trafficking is critical for establishing polarity, and therefore vectorial function, misfolded proteins are detected by the cell's exquisite quality control system and are trafficked to degradation pathways such as the proteasomes.

The crux of intestinal vectorial transport rests in the asymmetric distribution of the Na^+/K^+ ATPase pump to the basolateral membrane. As noted in Chapter 5, the β subunit of the pump is critical for sorting. By extruding three Na^+ ions in exchange for two K^+ ions, at the expenditure of ATP, the pump maintains a low (10 mM)$[Na^+]_i$ as compared to Na^+ concentrations of 140 mM in the plasma (Figure 2). In contrast, intracellular K^+ concentrations are in the 90–135 mM range with plasma K^+ at 5 mM. Critical to the effective functioning of the pump are mechanisms for the exit of K^+. The membranes are more permeable to K^+ than Na^+, and therefore, K^+ diffuses out of the cell more readily than Na^+ can enter the cell. These membrane transport properties and the fact that many intracellular proteins have fixed negative charges result in the interior of the cell having a negative potential as compared to either the luminal or the subepithelial compartment (Figure 2). Thus, the potential difference across the apical membrane (lumen to cell interior) can be distinct from that across the basolateral membrane (cell interior to the subepithelial space) and both contribute to the transepithelial potential difference. In intestinal preparations, transepithelial potential difference can be measured across the mucosa (which includes the subepithelial elements but not the longitudinal and circular muscle layers, and often referred to as stripped epithelial preparations) or across the entire gastrointestinal tract (including muscle layers). The cellular electronegativity along with low $[Na^+]_i$ establishes a strong electrochemical gradient for the passive entry of Na^+ into the cells. Epithelial cells take full advantage of this driving force to transport a variety of cations, including Na^+, anions, nutrients, and vitamins.

The above elements of polarity and asymmetric location of the Na^+/K^+ pump form the linchpin of epithelial fluid transport. In the intestine, these essential features are further nuanced to fit the specific function of the small and large intestinal segment, and even within an intestinal segment, there are regional differences between surface and villus cells and the crypt regions. The differences could be structurally decipherable such as subcellular localization or the simple presence or absence of essential transporters to differences in regulation of these proteins via posttranslational modifications or protein/protein interactions [114].

2.2 TIGHT JUNCTIONS

Our knowledge of cell–cell junctions, especially TJs, and the influence of the paracellular space on transepithelial ion transport has rapidly evolved in the past quarter century, from one of rather

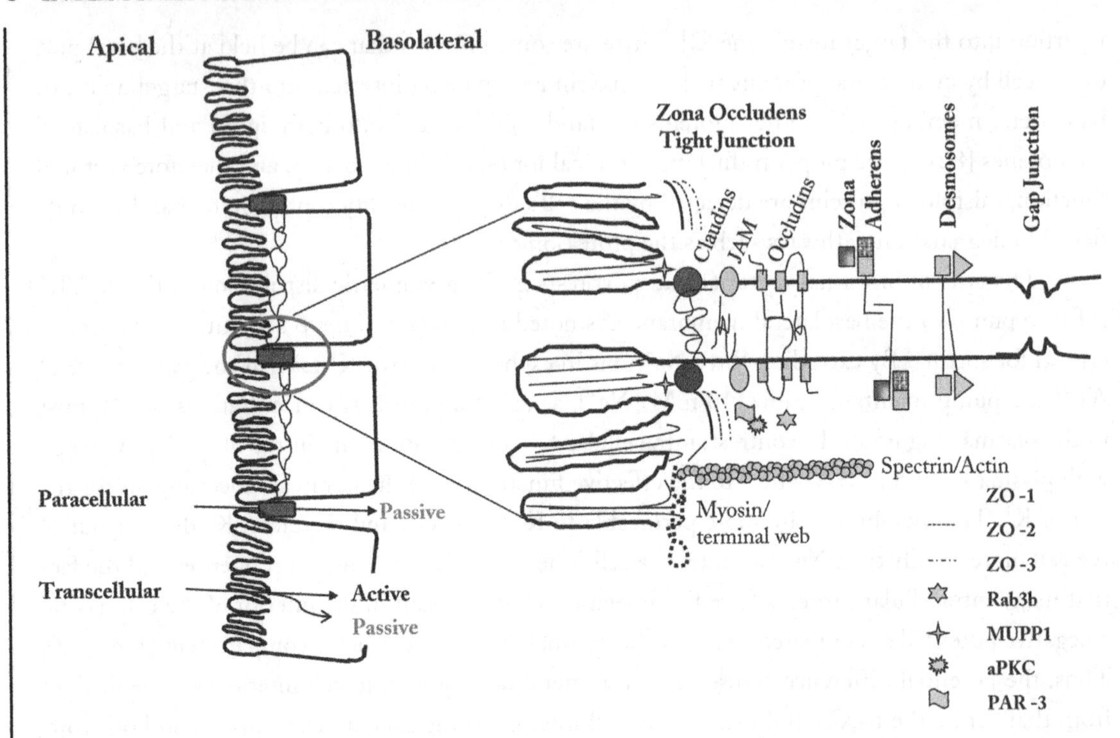

FIGURE 3: Intercellular connections. From the apical to the basal surface, there are the following cell junction types: zona occludens (ZO, tight junction), zona adherens (ZA), desmosomes and gap junctions. Tight junction is the most critical type in maintaining the cell and tissue polarity. Modified from Ref. 114 (Sleisenger and Fordtran's Gastrointestinal and Liver Disease, Chapter 99; permission obtained).

static structures to that of an exquisitely regulated, complex structure involving membrane spanning proteins, membrane-associated scaffolding proteins, and their intricate interactions (Figure 3). Together, the junctional complexes between the cells and the paracellular space delineate the epithelial barrier. This barrier function is conventionally measured as TER (R_t, in $\Omega \cdot cm^2$) or its reciprocal, conductance (G_s, in mS cm^{-2}). Epithelia with low transepithelial voltage (PD) have low TER and are classified as leaky, whereas those with a high transepithelial voltage have a high TER and are classified as tight. In comparison with other tissues, intestinal epithelia exhibit leaky to intermediate resistances. Thus, the jejunum is a leaky epithelium capable of transporting large amounts of fluid isotonically, a property that suits its function very well. In contrast, the distal colon shows higher resistance and is capable of reabsorbing water and electrolytes against an osmotic gradient. The overall trend is for an increase in TER down the cephalocaudal axis. Similarly, there is an increase in TER from the crypt regions to the villus [28]. Movement of water, electrolytes, and solutes across the tight junction are strictly passive and occurs in response to concentration, osmotic, electrical, and hydrostatic driving forces; nevertheless, the movement is influenced by the ability of the lateral space structure to undergo acute and long-term regulation, by electrical conductivity, and by size (≈7–15 Å radius) and charge selectivity lining the junctional spaces. For example, the permeability of Na^+ is 30-fold that of Cl^-, and size selectivity in the paracellular pathway can be regulated to permit the transmigration of leukocytes. The importance of the paracellular pathway and barrier function is underscored by its aberrations in various disease states.

Microscopically, several discrete structures contribute to the molecular fence as well as the cell to cell communications and paracellular geometry along the lateral walls of adjoining epithelial cells (Figure 3). Moving from the apical to the basal surface, the first is the zona occludens (ZO, tight junction, TJ), followed by the zona adherens (ZA), desmosomes, and gap junctions. Hemidesmosomes often contribute to the tethering of the epithelial cell to the basement membrane. Being the most critical, considerable attention has been focused on the physical structure, the component proteins, and the properties of the ZO region. As mentioned above, the TJ acts as a molecular fence protecting the asymmetry of apical and basolateral membrane composition and as a gate for transit through the paracellular space. As many as 50 different proteins contribute to the ZO, including claudins, ZO-1 proteins and occludins. Although not the first ZO proteins to be identified, the discovery of claudins, a family of 24 transmembrane cell adhesion proteins in 1998, was seminal in our understanding of TJs. Claudins exhibit homotypic and heterotypic interactions, and multiple claudins together can contribute to the overall barrier function. The claudins are 24–27 kDa in size, with four membrane spanning domains, two well-formed extracellular loops, and one intracellular loop. The first extracellular loop is ≈50 amino acids long and interacts with its counterparts on claudins (i.e., homotypic adhesion) of adjoining cells to form pores. The claudins lining the pores are important in determining charge and perhaps size selectivity of the TJ as there is evidence for at least two classes of pore sizes [3, 73, 104]. Thus, overexpression of claudin 2 does not affect the paracellular transit of large molecules but increases cation conductance and transit of small uncharged molecules. The second extracellular loop is 16–33 amino acids long and its function is unclear. The protein has a short, conserved N-terminal and a longer (21–63 amino acids), less well-conserved C-terminal cytoplasmic tail. The C-terminal tail is characterized by a three-amino acid terminal PDZ (PSD95, DIgA, ZO-1) binding motif, and overall, the C-terminal is believed to be the site of functional regulation. The first tight junctional proteins to be identified were the ZO protein family (e.g., ZO-1, ZO-2, ZO-3), but these are peripheral proteins. A hallmark of ZO-1 proteins are the three PDZ domains in the N-terminal region, which indicates their role in protein–protein interactions. Thus, the most amino terminal PDZ of the ZO proteins binds to the PDZ motif of claudins, and this interaction is necessary for proper targeting of claudin to the TJ.

The occludins are distinct from the claudins, but are also tetraspanning membrane proteins (60 kDa). Although the occludins were the second group of TJ associated and the first junctional transmembrane proteins to be identified, their function in the intestine is as yet unclear. Thus, while occludin knockout mice show complex manifestations, including autosomal recessive hearing loss, their intestinal transport function is unaffected. Other proteins in the TJ include junctional adhesion molecules (JAMs), a 43-kDa single-span transmembrane glycosylated protein with an elaborate extracellular domain and proteins such as multi-PDZ domain protein 1 and coxsackievirus and adenovirus receptor, which, along with the ZO proteins, contribute to the protein scaffold. All the transmembrane proteins have a cytoplasmic C-terminal tail with a PDZ domain that allows interaction with the scaffolding proteins. The scaffolding proteins are critical for the overall regulation of the fence and gate function of the epithelia. Through protein–protein interactions, they accomplish the following: they link to the cytoskeleton including a connection via actin filaments to myosin in the terminal web; they participate in vesicular transport involving the monomeric GTPase of the Ras superfamily (Rab3b); and they are involved in bringing to the proximity of the junctional complexes a variety of posttranslational signaling kinases and phosphatases that regulate junction assembly (for example, partition-defective proteins, PAR-3 and PAR-6, and atypical protein kinase C (PKC) [15, 16, 73]. These interactions, while stabilizing the scaffold and allowing for considerable cross-talk among regulatory signals, are also believed to be engaged in dynamic exchange and ultimately all of these influence paracellular permeability. The examples of how perturbation of the TJ proteins affect epithelial function are many and will be used through the text. A few examples are provided here: claudin 16 and claudin 19 are involved in a lumen-positive potential-driven paracellular Mg^{2+} and Ca^{2+} movement in the thick ascending limb of the loop of Henle in the kidney. Mutations in claudin 16 prevent the transit of Mg^{2+} and Ca^{2+}, whereas mutations in claudin 19 result in an increase in Cl^- permeability across the TJ from the serosa to the lumen, thereby nullifying the positive gradient and preventing absorption. Mutations in ZO-2 result in hypercholanemia and are associated with a poor targeting of claudins to the TJ and thereby disrupting its function. Finally, enteropathogenic *E. coli* alter TJ permeability by specifically activating PKC-ζ [112].

A critical player in the TJ mesh, via the scaffold proteins, are the actin and myosin filaments of the terminal web, which underlie the microvilli (Figure 3). Contraction of the perijunctional actomyosin ring has been associated with a number of physiological processes in the intestinal epithelium. The epithelium contains isoforms of myosin that are distinctive from those found in excitable muscle cells. A key step is the phosphorylation of the myosin II light chain (MLC), which is necessary for actomyosin contraction and results in the changes in barrier function associated with nutrient and ion transport in health and disease. Phosphorylation of MLC is catalyzed by members of the Ca^{2+}-calmodulin, MLC kinase family. Epithelial and endothelial cells predominantly express a 225-kDa isoform that is referred to as the long MLCK1. This protein is restricted to the perijunctional actomyosin ring of the villus enterocytes, and through a series of elegant studies involving specific inhibitors, antibodies, and targeted genetic manipulations, the central role of this kinase and MLC phosphorylation in regulating TJ permeability has been established by Marchiando et al. [73]. Thus, while MLCK activity and junctional permeability are increased during normal Na^+–glucose absorption, they are also increased in patients with IBD where it is associated with a loss of barrier function.

The next junctional complex down the lateral membrane and involved in intercellular interaction is the ZA. The major proteins involved in the epithelia are the E-cadherins; these transmembrane, 120-kDa glycoproteins have extracellular motifs, and the motifs of adjoining cells exhibit calcium-dependent homotypic interactions. These interactions contribute to cell-to-cell attachment and maintenance of cell polarity. Similar to the claudins, within the cell, the E-cadherins bind to specific proteins, the catenins, which are linked to the actin cytoskeleton through an additional family of actin-binding proteins, including radixin, vinculin, and a-actinin, and thereby anchor the junctional protein to the filamentous network of the cytosol. Signaling molecules, including the small G proteins, rab, and the tyrosine kinases src and yes, are associated with the ZA and link it to other intracellular regulatory cascades. Changes in catenin interaction with cadherin have been reported to occur during carcinogenesis [36]. Another junctional complex that is structurally similar to the ZA are the (hemi)desmosomes. In contrast to the ZA, instead of actin, the cadherin-like proteins of the hemidesmosomes anchor to intermediate filaments through a dense plaque of intracellular anchor proteins. Hemidesmosomes are present on the lateral and basal membranes of epithelial cells. On the lateral membrane, when they interact with their counterparts on the apposing cell membrane, they form a desmosome, or on the basal membrane, they help anchor the cell to the basement membrane.

Finally, junctional structures that have a radically different function than all those discussed thus far are the gap junctions; rather than buttress the gaps between cells, they bridge gaps between cells to allow neighboring cells to exchange small molecules. These junctions are made up of an assembly of connexins, a four-pass membrane spanning protein, and six connexins join to form a hemichannel. Alignment of two hemichannels from adjoining cells results in the formation of a continuous channel and allows the passage of small ions and molecules from one cell to the other. Thus, cholinergic activation of one epithelial cell, can trigger Ca^{2+} spikes in that cell and in cells further removed from that site in the same epithelium. Gap junctions have been extensively studied in the heart where they are critical for rhythmic contraction [2, 36].

• • • •

CHAPTER 3

Principles of Transepithelial Electrolyte and Water Movement

Transepithelial electrolyte, solute, and water movement can occur either from the lumen to the blood-side or the reverse and occurs either through the epithelial cell (transcellular) or extracellularly via the paracellular spaces (Figure 3). A variety of methodological approaches including in vivo triple lumen perfusion studies conducted in human subjects, animal models using intestinal loop studies, ex vivo studies using intestinal epithelial or membrane vesicle preparations, and reductionist models of intestinal epithelial cells in culture have contributed to our present understanding. The simpler models of studying transport in isolated apical and basolateral membrane vesicles or cells in culture or tissues in vitro have been invaluable in deciphering the molecular and cellular basis of electrolyte and water transport. However, the intact intestine is far more complicated and a number of features have to be taken into account when assessing the function of the intestine. Thus, factors that influence the movement of nutrients and solutes from the bulk luminal solution to the intestinal epithelium include the following: geometry of the intestinal wall; the influence of motility patterns including peristalsis and segmentation, villar motility controlled by the muscularis mucosa, and the finer contractions of the microvilli; and the glycocalyx, made of the extracellular glycosylated domains of apical membrane proteins, which contributes to the thickness and permeability of the unstirred layer. The unstirred layer can create a considerable diffusive barrier to the movement of lipophilic molecules through an aqueous milieu.

The information obtained from in vitro or reductionist models has to be confirmed in more complex in vivo models, cognizant of species and tissue differences; even so, at times different approaches may yield confounding observations. For example, the conclusion of some in vitro studies using tissues from the jejunum of CF patients was that there was an increased absorption of water due to a combination of increased Na^+ absorption and decreased Cl^- secretion. However, in vivo studies demonstrated that both passive Cl^- absorption and Cl^- secretion are decreased in the intestines of CF patients, implying that the severity of the disease is associated with decreased fluid absorption [95]. Clearly, the veracity of the findings is strengthened when multiple approaches yield similar and explicable results.

3.1 TRANSCELLULAR MOVEMENT

Transcellular transport can involve both passive and energy-dependent, i.e., active, transport mechanisms, whereas paracellular transport is strictly passive. Unidirectional fluxes of solutes can be measured from the mucosal to serosal solution (J_{ms}) or from the serosal to mucosal solution (J_{sm}). If the fluxes are measured in the absence of an electrical, chemical, or hydrostatic gradient, the general presumption is that cellular processes are being assessed. Net transport of a substrate is deemed "absorptive" if $J_{ms} > J_{sm}$ and "secretory" if $J_{sm} > J_{ms}$. In many epithelia, especially the intestine, the movement of fluid from the lumen to the blood is generally driven by Na^+ absorption, whereas the movement of fluid in the opposite direction is generally driven by Cl^- secretion. It is conceivable that net fluid accumulation in the lumen can occur simply by a decrease in absorptive fluxes with no change in J_{sm} as seen in some enteropathogenic *Escherichia coli*-induced diarrhea. On the other hand, pathogens such as *Vibrio cholera* cause copious amounts of fluid secretion by both inhibiting J_{ms} and increasing J_{sm}.

The cell interior is electronegative and thereby favors cation entry and anion exit from the cell. However, their transport across membranes requires specialized proteins, whereas the lipid plasma membrane is semipermeable and only nonpolar molecules, such as oxygen and CO_2, freely cross the lipid domain by simple diffusion (Figure 4a). Fat soluble vitamins and unconjugated bile acids can also cross by diffusion and the rate of transport is governed by the diffusion coefficients and concentration gradients. On the other hand, the transport of electrolytes, water-soluble nutrients, and larger lipophilic molecules require the presence of specialized transmembrane proteins. These could be channels (Figure 4b), facilitated diffusion carriers (Figure 4c and d), pumps (Figure 4e1), or secondary active transporters (Figure 4e2) and could be located on the apical or basolateral membranes. Other than the Na^+/K^+ ATPase located on the basolateral membrane of all intestinal epithelia, the distribution of other transporters is clearly region-specific and asymmetric.

The most rapid transit (>10^6 ions/sec) is via channels, composed of monomeric or multimeric transmembrane proteins that form a pore. The movements of ions through channels occurs by diffusion and is guided by their electrochemical potential and can be regulated by the structural properties of the channel. Thus, although extracellular concentration of Cl^- is high (≈110 mM) and intracellular is low (≈35 mM), the electronegative cell interior is sufficient for Cl^- to be close to or slightly above electrochemical equilibrium in the resting cell. When intracellular Cl^- increases, as seen in Cl^- secretion, the electrochemical driving force is sufficient for Cl^- to exit the cell against an almost 3-fold concentration gradient. Channels are defined by their ion selectivity and their current–voltage (I–V) relationship with the slope of the I–V curve representing conductance as dictated by Ohm's law. The ion selectivity is governed by the hydration radius of the ion and the physicochemical properties of the channel. Thus although K^+ ions are smaller and are also monovalent, they cannot be transported through some Na^+ channels. In addition to the number of channels in the membrane, the gating (time it remains open or close) of the channel defines its function.

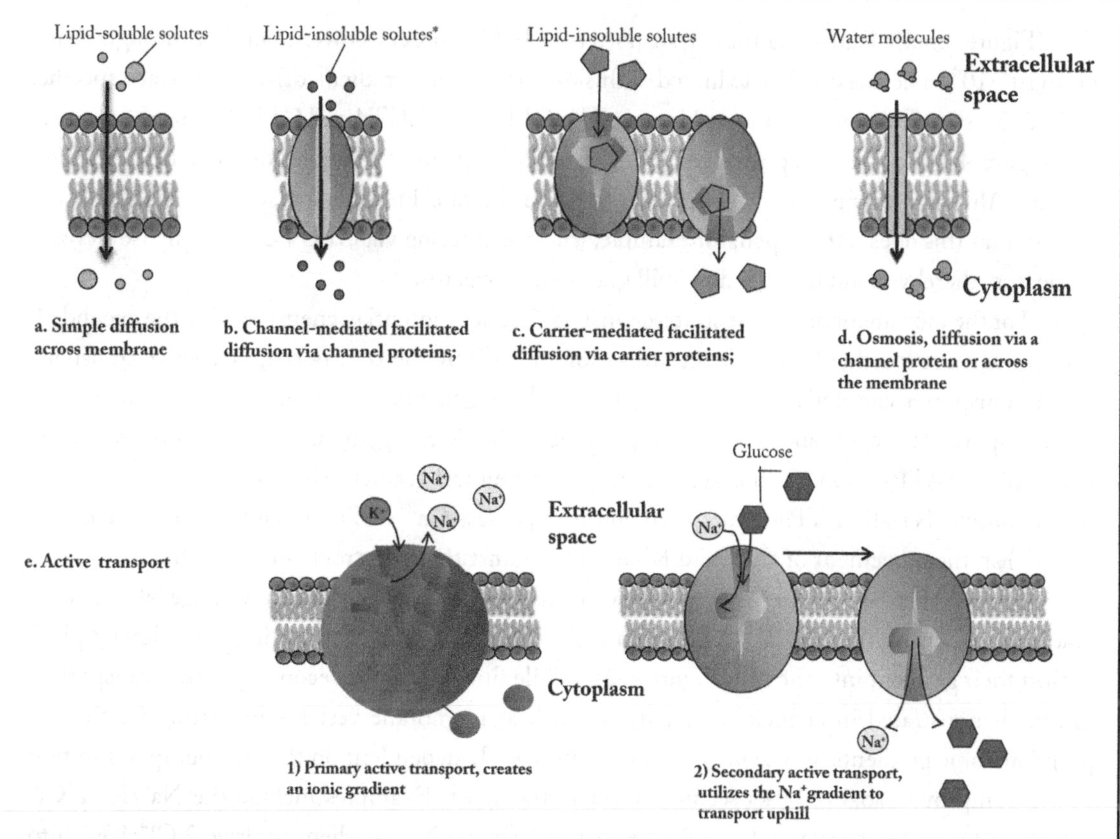

FIGURE 4: Channels, carriers and pumps. (a) By simple diffusion, nonpolar molecules freely cross the lipid domain. (b) Ion-specific channels mediate membrane transport through passive electrodiffusion. (c) Through facilitated diffusion, carrier proteins transfer the solute across the membrane by undergoing a conformational change. (d) Transcellular transport of water molecules occurs via channels proteins or carrier proteins. (e) Active transport occurs against an electrochemical gradient and can be driven by ATP (primary active transport; e1) or an ionic gradient (secondary active transport; e2). Modified from Ref. 114 (Sleisenger and Fordtran's Gastrointestinal and Liver Disease, Chapter 99; permission obtained).

Gating can be modulated by intracellular regulation, ion concentration or voltage. Depending on their structure, the current–voltage relationship of channels may not be linear and exhibit rectification. While most channels do not require ATP to function, the Cl^- channel, cystic fibrosis transmembrane conductance regulator (CFTR) is an exception in that activation of the channel requires nucleotide binding as well as phosphorylation events. Clearly mutations in a channel can affect any of its properties ranging from open probability to ion selectivity.

Other proteins act as carriers to facilitate the diffusion, down a concentration gradient, of noncharged substrates and down electrochemical gradients of some charged substrates across the membrane by undergoing a sequential and cyclical conformational change, rather than acting as a pore (Figure 4b–d). Transit via these transporters is 3–4 log orders slower than that through channels (10^2–10^4 molecules/sec). Facilitated diffusion carriers are gradient driven, substrate specific, and exhibit saturation and inhibitory kinetics. The GLUT2, GLUT4 and GLUT5 sugar transporters (Figure 4c) and the aquaporins (AQPs) (Figure 4d) are good examples of facilitated diffusion carriers. Although dissipation of the gradient will be the rate-limiting step, cell metabolism often ensures that this does not happen; for example, fructose entering via GLUT 5 is rapidly isomerized to glucose, thereby maintaining a downhill gradient for fructose.

For the movement of any substrate against its chemical potential, energy needs to be expended, and this can occur either by the direct utilization of ATP by the transporter, i.e., active transport, or the transporter can derive its energy by the gradient generated by a pump, i.e., secondary active transport. Active transporters, or pumps, directly derive energy by the hydrolysis of adenosine triphosphate (ATP), to move ions against an electrochemical gradient (Figure 4e1). In addition to the ubiquitous Na^+/K^+ ATPase, intestinal epithelia possess Ca^{2+}-ATPases and K^+/H^+ ATPases to accomplish the movement of Ca^{2+} and K^+ and H^+ against their electrochemical gradients.

Finally, there are numerous secondary active transporters that take advantage of the ionic gradients set up by the Na^+/K^+ ATPase to drive the movement of other ions or solutes uphill, against their gradient into the cell (Figure 4e2). While function of the secondary active transporters can be demonstrated in artificial preparations, such as membrane vesicles, by setting up the appropriate ionic gradients, in the intact cell they are acutely dependent on the Na^+ pump; inhibition of the pump by ouabain blocks secondary active transport. Examples include the $Na^+/K^+/2\ Cl^-$ cotransporter on the basolateral membrane that utilizes the Na^+ gradient to drive 2 Cl^-:1 K^+ into the cell against their chemical potential and the sodium-dependent glucose transporter (SGLT), which utilizes the Na^+ gradient to drive glucose into the cell across the apical membrane against its concentration gradient. The secondary active transporters exhibit substrate specificity and are capable of transporting one or more substrates in the same or opposite directions. For example, SGLT1 is isomer-specific transporting D-glucose but not L-glucose. SGLT1 and the $Na^+/K^+/2\ Cl^-$ cotransporters are symporters, transporting substrates in the same direction. Other secondary active transporters such as the Na^+/H^+ exchangers, are antiporters, utilizing the Na^+ gradient to move two ions in opposite directions, viz., an H^+ out of the cell and a Na^+ into the cell.

3.2 PARACELLULAR MOVEMENT

As discussed in Chapter 2 paracellular movement of fluid, solutes, and electrolytes is driven by concentration, osmotic, electrical, and hydrostatic gradients. The relative permeability of the junctional complexes including charge selectivity, is dictated largely by the proteins of the TJs (Figure 3). Although there are differing conclusions on how significant the geometry of the lateral spaces are in modifying paracellular movement, they are nevertheless worthy of a mention. In addition to the glycocalyx and the TJs per se, transit through the lateral spaces can be modified by the other junctional complexes, such as the ZA and desmosomes. Local ionic gradients can also be created by the transporters on the basolateral membranes, and finally, for the transported fluid and electrolytes to reach the subepithelium, they must traverse the basement membrane. The movement of charged particles is governed by the electric potential, concentration differences, and charge selectivity of the TJ, whereas the movement of uncharged particles is dictated by concentration gradients and size selectivity of the TJs. A passive transport mechanism that is especially relevant when bulk flow of water occurs, as in response to a meal, is solvent drag. The water movement results in a nonspecific entraining of solutes and electrolytes; K^+ transport across the small intestine occurs via solvent drag as does some fraction of the glucose transport in response to a meal.

• • • •

CHAPTER 4

Intestinal Architecture and Electrolyte Transport

The overall architecture of the intestine including the muscle and epithelial layers is geared toward the optimal digestion of food and subsequent absorption of nutrients, water and electrolytes. These structures and underlying functions are also critical in maintaining the barrier function of the epithelium to prevent noxious luminal stimuli from entering the blood stream. The musculature can influence bulk fluid flow and alter the transit time for the chyme through changes in motility by any of the smooth muscle layers, longitudinal, circular, or muscularis mucosa. Specialized musculature, like the tenia coli in the colon, help in haustral contractions, which increase the exposure of the colonic epithelium to the hypertonic luminal contents, allowing sufficient time for the reabsorption of water and electrolytes. Hypermotility in either the small or large intestine can increase transit times, allowing for inadequate reabsorption and resulting in diarrhea. However, the bulk of the work associated with fluid transport rests with the epithelium. Here again, better exposure to luminal contents allows for more efficient absorption. The structure of the small intestine lends itself to a massive amplification of the surface area and therefore increased ability to absorb; contributing factors include the plicae circulares (circular folds of Kerckring; 3-fold), the finger-like projections of the villus–crypt architecture (10-fold), and the amplification of the apical membrane of the enterocyte into microvilli (20-fold). This computes to a ≈600-fold amplification of the surface area, and if the intestines were considered to be a simple cylinder, the epithelial surface would be 3300 cm^2 as compared with the actual surface area of 1,980,000 cm^2. While not as expansive as the small intestine, the large intestinal epithelial surface is also amplified, and the spatial separation of crypts and surface cells is needed for maximal reabsorption and key metabolites such as short-chain fatty acids (SCFAs) produced by the action of the microflora.

As described in the previous chapters, the intestine serves as the first line of defense between the mucosal compartment and the blood side. The properties of the apical membrane and the (im)permeability of the TJs contribute to this defense, and a loss of integrity of these barriers leads to a disruption of vectorial transport and access of toxic substances to the subepithelium and systemic circulation leading to disease. Intestinal epithelial cells create a luminal milieu that can deter microbial pathogen progress. Thus, interdigestive migratory myoelectric contractions and mucous secretions from the glandular cells attenuate microbial colonization and growth. Furthermore, the Paneth cells at the base of the crypt secrete small-molecular-weight (18–45 kDa) cysteine-rich compounds, the defensins. These cationic molecules are antimicrobial (bacteria, fungi, and some viruses) and act by binding to the microbial membrane, causing pores and allowing the leakage of ions and nutrients critical for microbial function. Another line of defense is the gut-associated lymphoid tissue (GALT), located in the subepithelial space with some specialized epithelial cells (such as the M cells). While GALT has some commonalities with the systemic immune system, it has many characteristics uniquely suited to its role of providing gut immunity.

4.1 SEGMENTAL HETEROGENEITY OF TRANSPORT DOWN THE CRYPT–VILLUS AND CRYPT–SURFACE AXES

The crypt–villus axis in the small intestine and the crypt–surface axis in the colon exhibit significant spatial heterogeneity with respect to cell structure and function, including that of transport proteins. The prevailing hypothesis, based on a considerable body of work, is that intestinal secretions largely occur in the crypts, whereas absorption occurs chiefly in the villus cells of the small intestine and in surface colonocytes, with a gradation of functions in between. The intestinal epithelium is comprised of four major types of cells: the majority are the "workhorse" enterocytes; endocrine cells and goblet cells are distributed along the crypt–villus axis, with the goblet cells being especially abundant in the colonic epithelium; and the Paneth cells, which produce defensins, are at the base of the crypt. All these cells arise from the epithelial stem cells, which reside in the proliferative zone near the base of the crypt. As the cells divide, a few of them migrate downward to form the Paneth cells, and the bulk of them migrate upward to eventually form villus enterocytes in the small intestine or surface colonocytes in the large intestine, in addition to the endocrine and goblet cells. The intestinal epithelium is a highly metabolic tissue and rapidly turns over; in the small intestine, the cells undergo apoptosis and slough after 3 to 5 days at the tip of the villus, whereas in the colon, the turnover is in the order of 3 to 8 days [64].

As the epithelial cells migrate, they undergo marked differentiation with visible changes in the development of a pronounced microvillar architecture, the underlying cytoskeleton, a greater abundance of mitochondria, and an increase in the TJ complexity. This is accompanied by a pronounced change in some of the signaling cascades, and an altered expression of some membrane proteins, including transporters [12, 86]. As depicted in Figure 5 and 6, some transport molecules are found at relatively constant concentrations along the axis, whereas some of them exhibit a greater density in the base of the crypt and others are denser toward the villus or surface. Thus, it is not surprising that the levels of the ubiquitous Na^+ pump remain relatively constant, while others, such as the signaling molecule adenylate cyclase and the cAMP-associated

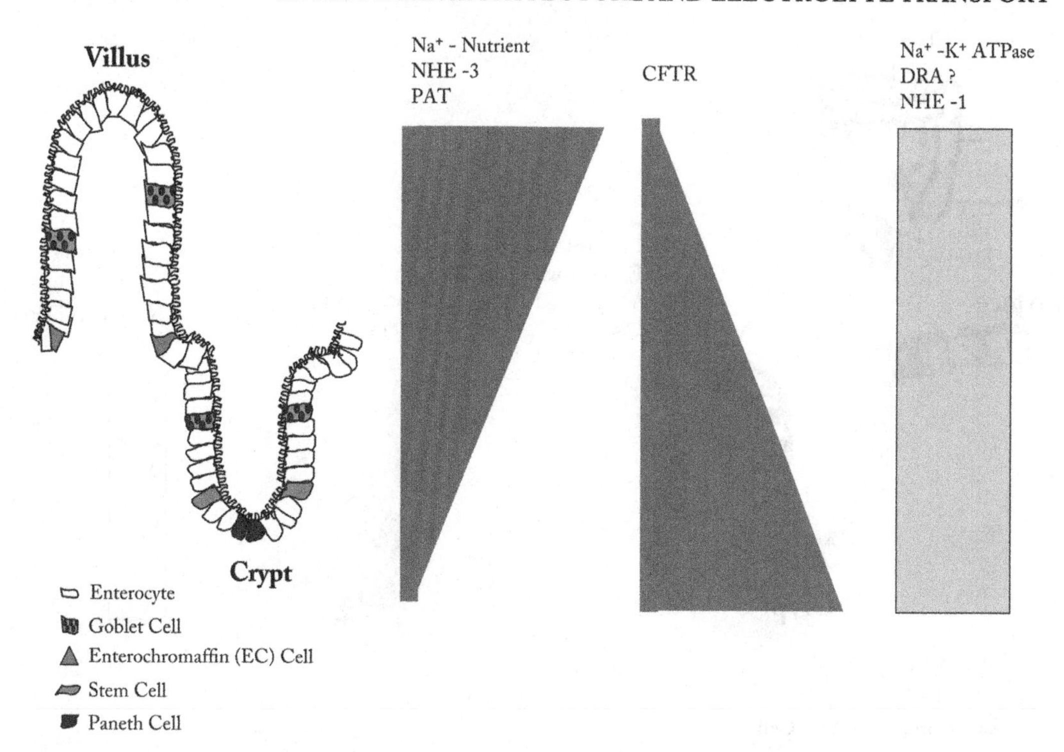

FIGURE 5: Transport protein gradients in the small intestine. Along the crypt–villus axis, transport proteins display a significant spatial geometry. Some proteins are distributed relatively evenly along the axis, while others exhibit a greater density at either end. NHE, sodium–hydrogen exchanger; PAT, putative anion transporter; CFTR, cystic fibrosis transmembrane regulator; DRA, down-regulated in adenoma. Modified from Ref. 114 (Sleisenger and Fordtran's Gastrointestinal and Liver Disease, Chapter 99; permission obtained).

Cl^- channel CFTR, decrease in more mature villus cells. In contrast, there is increased expression of Na^+ nutrient-coupled transporters, apical Na^+–H^+ exchangers, and brush border membrane hydrolases as the crypt cells differentiate into villus cells. In diseases such as IBD, celiac disease, and some enteric infections, there is a selective destruction of the surface or villar epithelium, and secretion predominates. These underscore the villus–crypt segregation of absorptive and secretory functions.

Like many biological phenomena, the dichotomy between secretory crypt cells and absorptive surface cells is not rigid, and examples of where colonic surface cells secrete Cl^- and conversely where colonic crypt cells absorb Na^+ have been reported [28, 32, 80]. The premise of this crypt–villus model has been questioned by Lucas et al. (pp. 478–484 of ref. [80]). Based on studies examining heat stable enterotoxin (STa), STa induced diarrhea using fluid recovery techniques. Other

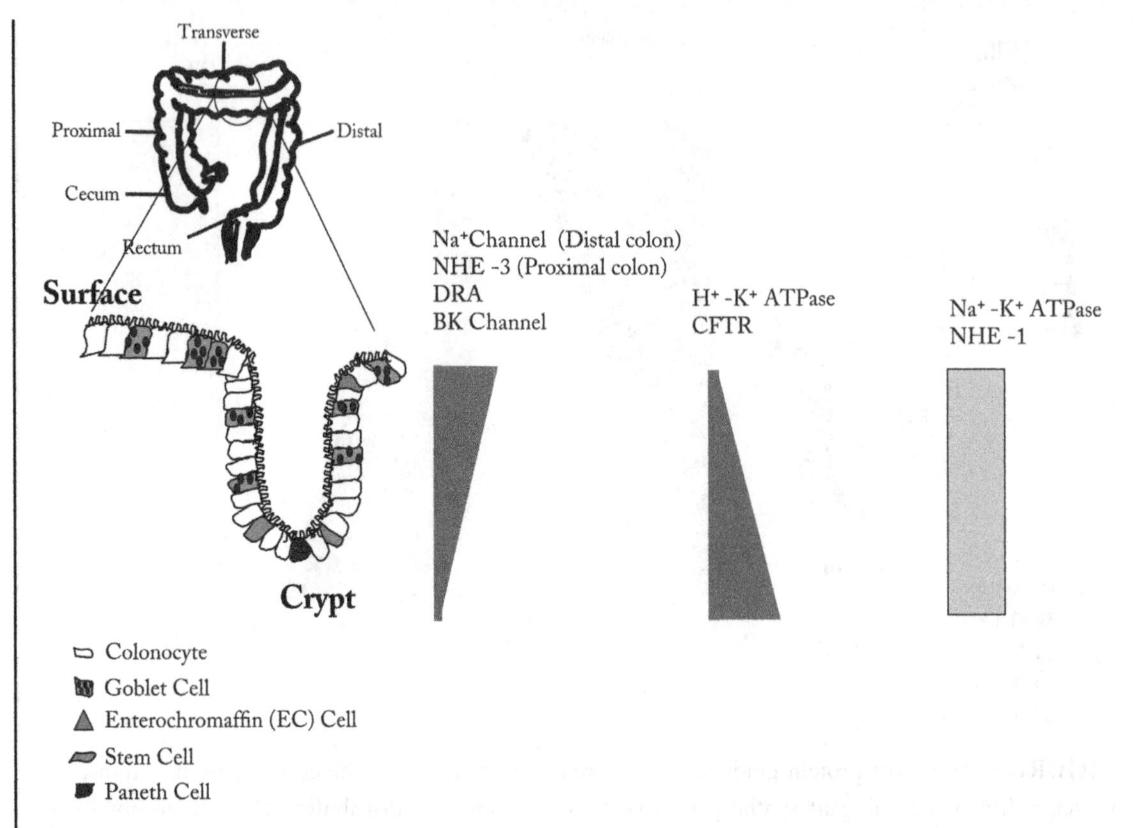

FIGURE 6: Transport protein gradients in the large intestine. Similar to small intestine, along the crypt: surface axis, transport proteins display a significant spatial geometry. NHE, sodium–hydrogen exchanger; CFTR, cystic fibrosis transmembrane regulator; DRA, down-regulated in adenoma. Modified from Ref. 114 (Sleisenger and Fordtran's Gastrointestinal and Liver Disease, Chapter 99; permission obtained).

explanations such as increased interstitial pressure, intracapillary pressure, and hydraulic conductivity of tight junction have been proposed. While these should be considered, Geibel et al have outlined compelling arguments in favor of the crypt/villus model [80]. Undoubtedly, the cross-talk between various signaling cascades and the transporters can vary in the crypt–villus axis and thereby fine-tune intestinal function.

4.2 SEGMENTAL HETEROGENEITY OF TRANSPORT DOWN THE CEPHALOCAUDAL AXIS

While the entire length of the intestine is geared toward transport of water and electrolytes, each segment of the intestine is geared toward distinct physiological functions. It is therefore not surprising that there is significant heterogeneity in the types of transport mechanisms and transporters encountered down the length of the cephalocaudal axis. The TER increases down the length of the cephalocaudal axis, with the duodenum being "leaky" with a TER of ≈15 to 20 $\Omega\cdot cm^2$ in the jejunum and 30 $\Omega\cdot cm^2$ in the distal ileum and the colon showing the highest TER of 50–60 $\Omega\cdot cm^2$.

The duodenum receives acidic and hypertonic chyme, and its bicarbonate-rich secretions are geared toward neutralizing the chyme and making it more isotonic. This is also helped by the alkaline pancreatobiliary secretions, and the process of digestion begins in this region.

The bulk of nutrient absorption occurs throughout the length of the jejunum with more lipophilic substances being taken up in the more distal parts of the jejunum. The jejunum is well suited for the absorption of copious amounts of water following nutrient absorption via the glucose- and amino acid-coupled transporters. The distal ileum has highly specialized transporters for the absorption of bile acids and vitamin B12. The steatorrhea observed upon ileal resection is due to bile acids and undigested fatty acids stimulating colonic Cl^- secretion. In the colon, including the cecum, proximal colon, and distal colon, fluid reabsorption needs to occur against a fairly steep osmotic gradient. The specialized motility patterns such as haustral contractions allow for effective kneading of the chyme and exposure to the epithelial surface. In addition, the colonic epithelia show segmental differences in Na^+ transporters—the cecum and proximal colon possess Na^+/H^+ exchange mechanisms, whereas the distal colon exhibits electrogenic Na^+ absorption, which is necessary for the final fluid extraction in preparation of the feces for excretion (43, 108–110). There is also a clear segmental heterogeneity with respect to K^+ transport (Figure 5 vs. 6). The distal colon possesses K^+/H^+ ATPase, which sequesters K^+. Colonic epithelia have apical channels to secret K^+ when luminal $[K^+]$ falls <25 mM. Although the segmental heterogeneity in the distribution of Na^+ channels and nutrient transports is teleologically sound, the reasons for the restricted distribution of other transporters is less obvious and remains to be elucidated. For example, apical membrane anion exchange processes are a critical feature of both small and large intestinal ion transport, yet the colon predominantly expresses the down-regulated in adenoma (DRA) protein, whereas the small intestine exhibits other distinct molecular entities such as the putative anion transporter (PAT-1) [50, 55, 71] (see Section 5.2.1).

• • • •

CHAPTER 5

Electrolyte Transporters—Pumps, Carriers, and Channels

5.1 TRANSPORT OF CATIONS

The bulk of cation transport across the intestinal epithelium is accounted for by Na^+ and K^+. Also of physiological import, is the transport of the divalent cations such as calcium and metal ions such as iron. Absorption of calcium and iron largely occur in the duodenum and proximal jejunum. The villus enterocytes possess elaborate carriers and sequestration mechanisms to ensure that while absorption is occurring maximally, the intracellular free concentrations of these ions are low and do not reach levels potentially toxic to the cell. The transport of sodium and potassium involves channels, secondary active transporters, and the linchpin, Na^+/K^+ ATPase pump. As mentioned in Chapter 4, while the pump is uniformly distributed other transporters are variably distributed both along the villus-crypt and the cephalocaudal axes (Figure 4 and 5). In the past three decades the advent of sophisticated molecular cloning technology, the invention of patch clamp methodology and of live cell imaging has vastly improved our understanding of the nature of these transporters and their coordinated functioning. A note about nomenclature—there are more than 360 identified solute carriers (SLC) which are classified, largely based on sequence homology into 46 families. Many of these transporters were traditionally classified with abbreviations based on function, e.g., NHE for Na^+/H^+ exchangers and AE for anion exchangers. The SLC classification for the NHE family is SLC9A; for example, NHE3 is also known as SLC9A3. For ease of reading both the SLC nomenclature and more common names are provided.

5.1.1 Sodium

Throughout the length of the intestine, the majority of cellular sodium transport processes are geared towards net transepithelial absorption of Na^+ (i.e., J_{ms} Na^+ is > J_{sm} Na^+). The movement of Na^+ that occurs during secretion, is mostly through the lateral spaces and tight junctions of the paracellular pathway to electrically balance the cellular anion (largely Cl^-) secretion. The underlying driving force for sodium absorption is the basolaterally located Na^+/K^+ ATPase which extrudes

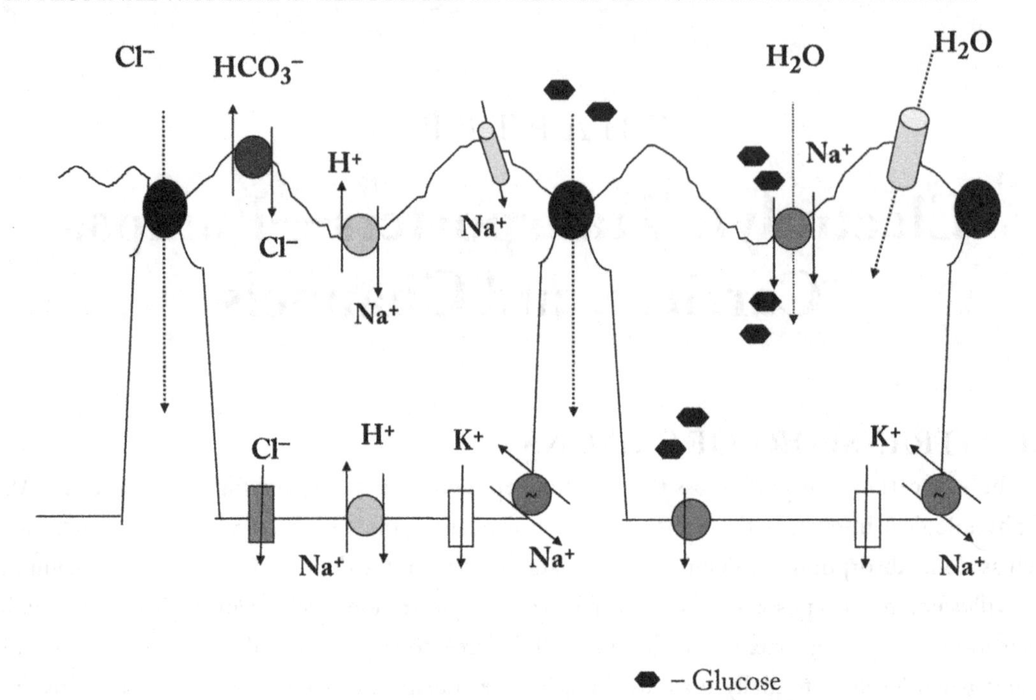

FIGURE 7: Sodium transport. Sodium crosses epithelial cell apical membrane down an electrochemical gradient. It can happen through three mechanisms: (1) an ion-specific channel that can be blocked by amiloride; (2) a carrier that couples the movement of sodium and nutrients, such as glucose (e.g., SGLT1); (3) a carrier that allows electroneutral entry of sodium in exchange for intracellular hydrogen (antiport carrier) (e.g., NHE-3). The common exit pathway across the basolateral membrane is the sodium pump. Modified from Ref. 114 (Sleisenger and Fordtran's Gastrointestinal and Liver Disease, Chapter 99; permission obtained).

$3Na^+$ for entry of $2K^+$ at the expenditure of one ATP. Depending on the segment of the intestine, a combination of antiporters, nutrient-dependent symporters and channels, utilize this gradient to transport Na^+ into the cell across the luminal membrane (Figure 7). Regardless of the mechanism of its entry, the Na^+ exit from the cell is by active transport via the Na^+ pump at the basolateral border [6]. In the proximal regions of the small intestine (especially jejunum) the bulk of sodium absorption is coupled to solute (nutrient: sugars and amino acids) (Figure 5) transport and stimulated by the presence of sugars. In the ileum and proximal colon, Na^+/H^+ exchangers predominate whereas in the distal colon (Figure 6), in most species, transport is via Na^+ channels. It has been suggested that in a normal adult mammal, solute-dependent cotransport accounts for as much as 80% of net Na^+ absorption, the remaining 20% occurring via antiporter or channel mediated transport.

The movement of Na^+ is coupled to that of Cl^-, and the mechanisms, which involve both cellular transporters and electrical coupling, vary down the length of the cephalocaudal axis. In the distal ileum and proximal colon, this coupling largely occurs through the coordinated actions of the Na^+/H^+ and the Cl^-/HCO_3^- exchangers, governed by cell pH and volume. Thus activation of NHE, results in the extrusion of H^+ and alkalinization of the cell, since cellular carbonic anhydrases, generate HCO_3^- and H^+. The HCO_3^- is extruded from the cell in exchange for Cl^-, via Cl^-/HCO_3^- exchangers (Figure 7, left cell, see Section 5.2). The net result is maintenance of cell pH, cell volume, electroneutral Na^+ absorption and extrusion of H_2O and CO_2 ($H^+ + HCO_3^-$). While Cl^-/HCO_3^- exchangers are present in the proximal small intestine and in the distal colon, their activity is less tightly linked to the activity of NHEs. In the jejunum, Cl^- movement could be electrically coupled to that of Na^+ wherein paracellular Cl^- transport, follows the transcellular movement of Na^+ across the epithelium to maintain electroneutrality. In the distal colon, Cl^-/HCO_3^- exchangers function in a Na^+-independent manner [6].

5.1.1.1 Sodium–Potassium ATPase. Since the Na^+/K^+ ATPase is the driving force behind most of the ions absorbed and secreted, it is imperative that this pump be discussed first. The Na^+/K^+ ATPases have been well characterized and, in most epithelia, with the exception of the choroid plexus, are localized at the basolateral membrane of the intestinal epithelium (Figures 2 and 7–9). The protein is a P-type ATPase whose turnover involves the hydrolysis of ATP and covalent phosphorylation of an aspartate residue in the active site, resulting in conformational changes; dephosphorylation events eventually restore the protein to its initial conformation. The net result is the extrusion of 3 Na^+ from the cytoplasmic to the extracellular side and pumping in 2 K^+ for every molecule of ATP hydrolyzed.

The Na^+/K^+ ATPases have three subunits, α β and γ with a 1:1:1 stoichiometry in the membrane. Although the α subunit carries the essential elements for catalysis, viz., the Na^+, K^+, and ATP binding sites, and translocates the ions, the β subunit is essential for proper membrane insertion of the α subunit. In addition, both the β and the γ subunits influence the affinity of Na^+/K^+ ATPase for Na^+, K^+, and ATP. The α subunit is a multipass, ~113-kDa transmembrane protein with four isoforms, with the α_1 isoform being ubiquitously expressed. The α_4 isoform is primarily expressed in the brain, lung, adipose tissue, and various striated and smooth muscle cells; the α_3 is expressed in the heart and neuronal cells; and α_4 is testis-specific. The N- and C-terminal regions of α are cytoplasmic, and the subunit has 10 transmembrane helices, functionally and structurally separated in two segments and a number of key cytoplasmic domains. The membrane helices M1–M6 represent the transmembrane core, which occurs in a number of P-type ATPases, whereas the M7–M10 segment is specific for the Na^+/K^+ ATPase and H^+/K^+ ATPases and influences Na^+ affinity; together, the transmembrane helices contribute to the ion-binding sites. The cytoplasmic domains include two A (or actuator) domains that sandwich the M1 and M2 transmembrane helices, phosphorylation (P) and nucleotide (N) binding domains that separate the M3–M4 helices from the M5–M6 helices and a domain C terminal to the M7–M10 helices. The α subunit also has the binding site for cardiac glycosides, such as ouabain, potent inhibitors of the pump, and the A, P, and N sites are necessary to mediate cation transport [78].

The β subunit is a 45-kDa, single-pass transmembrane glycosylated protein, with a very short cytoplasmic N terminal and a large heavily glycosylated C-terminal ectodomain; it has three isoforms and is necessary for protein trafficking to the appropriate membrane as well as modulating the affinity for Na^+ and K^+. The β_1 isoform is ubiquitously expressed, whereas β_2 is primarily expressed in neurons, muscles and heart cells and β_3 is found in testis, lung and liver. Finally, the γ subunit, also termed FXYD2 because of its signature sequence, is a single-pass transmembrane protein of 61 amino acids that modulates the affinity of the pump for Na^+, K^+, and ATP. There are six other members of the FXYD family, many of which are associated with the pump but are not termed the γ subunit. One of these, FXYD4, may play a special role in the response of the distal colon to mineralocorticoids. While aldosterone induces Na^+/K^+ ATPase expression and activity in many tissues, in the distal colon, it does so indirectly through the up-regulation of corticosteroid-induced factor (FXYD4). Both the β and γ subunits associate with the M7–M10 domain of the α subunit to modulate function, and the α subunits association with M5 is thought to be important for K^+ affinity. In addition to the three subunits, recent studies show interactions with other proteins, such as the Na^+/K^+ ATPase interacting proteins, which may further influence the functioning of the pump [78].

Other pumps involved in intestinal cation transport are the H^+/K^+ ATPases located on the apical membrane of colonocytes and involved in K^+ absorption and the Ca^{2+} ATPases on the basolateral membrane, involved in Ca^{2+} absorption. Like the Na^+/K^+ ATPases the turnover of these proteins also involves a phosphorylated intermediate.

The cells of the intestine, utilize the chemical gradient set up by the Na^+/K^+ pump to transport a host of solutes including essential nutrients, such as glucose and amino acids, and bile acids into the cell by secondary active transport (see Section 3.1). Some examples are discussed below.

5.1.1.2 Sodium-Coupled Glucose Transport. The quintessential, sodium-coupled solute transporter (SLC), is the sodium-dependent glucose transporter, SGLT, a symporter belonging to the *SLC5A* gene family. The SLCs constitute a major group of proteins, and many of these transporters use energy that is stored in the sodium gradients to drive a variety of nutrients. The two most well-known members belonging to the *SLC5A* gene family are the *SGLT1* found in the small intestinal mucosa and *SGLT2* chiefly located in the renal proximal tubule. In the intestine glucose enters the absorptive cell by SGLT1 located on the brush border membrane and exits into the blood by GLUT2, a member of the facilitative glucose transporter family, on the basolateral membrane [46] (Figure 7, right cell).

Transport across SGLT1 is electrogenic (2 Na^+:1 glucose), substrate-specific (transports galactose but not fructose) and stereospecific (D- and not L-isomer). Fructose enters the cell by the facilitated diffusion carrier GLUT5 (glucose transporter 5), which is highly specific for fructose, and exits by GLUT2; thus, glucose and fructose share a common exit pathway. Cotransport of Na^+ and glucose via SGLT is completely reversible. The direction of transport depends on the direction of the sugar gradient and the Na^+ electrochemical gradient. The kinetics for SGLT can be explained as follows: Two external Na^+ binds first to SGLT1 and increases the affinity of the transporter for sugar, which then changes the conformation to deliver the sugar and Na^+ to the other side of the membrane. The release of Na^+ and sugar causes another conformational change to expose the sugar and Na^+ binding site to the external environment to continue the process of absorption [127]. Crystal structure of a member of the sodium solute symporter, the Na^+/galactose transporter [26], revealed a structure that contains 14 transmembrane helices with repeats of 5 helices (TM2–6 and TM7–11) in an inward-facing state. This structure was similar to a different gene family, that of the leucine transporter. The current thinking is that cotransport with amino acids or with bile acids (primarily in the ileum) proceeds by a similar mechanism.

SGLT has also been demonstrated to serve as a water channel that transports 210 water: 2 Na^+:1 glucose [126] and therefore may account for up to 5 L of fluid reabsorption following a meal. The SGLT1 also behaves as channels for small hydrophilic solutes when expressed in oocytes, and this osmotic permeability is blocked by the SGLT1 inhibitor phlorizin. It is interesting to note that intestinal brush border lacks functional AQPs, and thus SGLT1 can play an important role in fluid absorption in this epithelium. The relevance of this function is highlighted by the role *SGLT1* plays in oral rehydration therapy (ORT), which saves thousands of lives annually. ORT (Section 8.6) is based on the stimulation of fluid absorption by sugar and is used to treat severe diarrhea. Although not widely prevalent, the inherited disorder glucose–galactose malabsorption, which manifests as diarrhea, is due to mutations in *SGLT1*.

In response to a meal, not all glucose absorption occurs by secondary active transport via SGLT1. There is increasing evidence for both a facilitated diffusion component and a paracellular component of glucose absorption. In the former, glucose absorption is mediated by the glucose- and hormone-dependent regulation of GLUT2 at the brush border membrane [48, 49]. The rapid trafficking of GLUT2 to the brush border membrane is controlled by the SGLT1-dependent activation of a PKC-dependent pathway and also by MAP kinase intracellular signaling pathways. As mentioned in Section 2.2, through the activation of the terminal web MLCK, there is increased junctional permeability and glucose transit in response to a meal [73]. Thus, the intestine is able to deal efficiently with the luminal glucose load associated with a meal.

5.1.1.3 Sodium–Hydrogen Exchanger. Sodium absorption has been studied extensively for many years in both in vivo and in vitro preparations using molecular and physiological approaches. Electroneutral sodium absorption can occur by the activity of one or more of the 10 members of the Na^+/H^+ exchanger family (NHEs or *Slc9a*). The steep extracellular to intracellular Na^+ gradient established by the Na^+/K^+ ATPase pump dictates NHE activity, whereby the extracellular Na^+ is exchanged for intracellular H^+ (Figure 7, left cell). Overall members of this family of proteins play a crucial role in fine-tuning intracellular pH, plasma electrolyte, and volume homeostasis and cell volume control.

There are cell and segmental differences in the distribution of the NHE isoforms. The NHE-1 isoform is almost always expressed on the basolateral membrane and is involved in "housekeeping" functions of cell volume regulation and growth. In the ileum, NHEs (2, 3 and 8) on the brush border membrane, linked to Cl^-/HCO_3^- exchangers make up the neutral NaCl absorptive process, accounting for most of the Na^+ absorption in this segment. In the proximal colon, there is involvement of apical membrane NHE activity (NHE-2, -3 and -8), is linked to Cl^-/HCO_3^- exchangers, as well as to $SCFA^-/OH^-$ exchangers. However, in the distal colon, the epithelial Na^+ channel (ENaC) plus elevated tight junction resistance accounts for high efficiency Na^+ absorption and prevents the body from becoming Na^+-depleted even when on a virtually Na^+-free diet.

Ten isoforms of NHEs (SLC9A1–10) have been isolated and cloned, and except NHE5 and NHE10, all the other NHE isoforms have been detected in different segment of the gastrointestinal tract. The NHEs are 12-pass transmembrane proteins (645–898 amino acids in length) which are structurally and functionally related. The proteins have a short N-terminal cytoplasmic sequence followed by 12 transmembrane helices and a long C-terminal cytoplasmic domain. Not surprisingly the proteins are fairly well-conserved (50%–60% homology) in the N-terminal and membrane domains, the latter of which is involved in the ion transport function. In contrast, the C-terminal domains are involved in regulation and exhibit only a 20%–30% homology of the amino acid sequence of the isoforms [44]. Four of the NHE isoforms are identified in the plasma membranes of intestinal enterocytes; they are NHE 1–3 and NHE 8 (SLC9A1–3, SLC9A8). The isoform first identified in the intestine, NHE1 (initially described as the growth factor-activatable NHE), is present only on the basolateral membrane of enterocytes, is involved in HCO_3^- secretion and does not play a direct role in absorption. Studies thus far in the human intestine show that the expression of NHE1 is relatively uniform, with no differences in the expression of NHE along the crypt–villus axis or along the longitudinal axis suiting its definition as the "housekeeping isoform" [25].

NHE2, NHE3 and the recently characterized NHE8 are found in both the small intestine and the colon, in the apical membrane of surface cells, and are thought to play a role in vectorial movement of Na^+ across the epithelium. Evidence to date suggests that of these, NHE-3 is perhaps the critical transporter for intestinal Na^+-dependent fluid absorption in the adult and is briefly reviewed here. In terms of tissue distribution, in the human, NHE2 is expressed in all intestinal epithelia with highest expression in the proximal colon, whereas mRNA expression of NHE3 appears to be highest in the jejunum and colon, with no difference between the distal and the proximal colon [51]. There is a controversy regarding the expression of NHE2 with it demonstrated to be expressed primarily in the absorptive villus of the rat intestine [13], while other studies in the mouse colon that report their expression in the crypt base cells [38]. However, studies in human colonic biopsies showed uniform distribution of NHE2 along the crypt/surface axis [25]. Further, in isolated colonocytes from the crypts, wild type and NHE2- null mice had similar NHE activity, suggesting that the loss of NHE2 did not have an effect on absorption [38]. In fact, these studies report a compensatory 2-fold increase in NHE3 mRNA. In contrast, NHE3 can be considered as a marker for absorptive epithelia, as they are expressed only in the apical membranes of villus or surface cells but not in crypts [51]. Studies on NHE2 knockout mice demonstrate that NHE2-deficient mice suffer from gastric dysfunction without experiencing intestinal disability, whereas the NHE3 knockout mouse suffers from diarrhea and mild metabolic acidosis. NHE8 is located in the apical membrane and helps with absorption in the early stages of development in suckling rats at a time when NHE2 and 3 are absent [129]. NHE8 levels decreases as the intestine matures and this inhibition has been shown to be regulated by EGF [128]. Taken together, these results suggest a more important role for NHE3 in Na^+-dependent intestinal fluid absorption in surface cells in the adult [106].

Although not in the intestine, the NHE4 isoform is expressed primarily in the basolateral membrane of gastric parietal and chief cells and coupled with a basolateral Cl^-/HCO_3^- exchanger (AE2) plays an important role in the regulation of gastric acid secretion. NHE5 is expressed in the brain and NHE10 is expressed in osteoclasts and in sperm. NHE6, -7, and -9 appear to be on intracellular membranes, with NHE6 and NHE9 being localized to recycling endosomes and NHE7 to the trans Golgi network. In addition to these 10 isoforms, the presence of a novel Cl^--dependent NHE isoform, both molecularly and functionally, was reported in the apical membrane of rat distal colonic crypt cells [100]; however this remains to be confirmed [5].

The molecular identification of isoforms of NHE permits a more thorough understanding of the basis for the differing responses observed in NHE activity to a variety of agonists. A number of studies have centered on the regulation of NHE3 activity, as this isoform appears to be the one most subject to hormonal manipulation. Regulation could involve protein phosphorylation, protein-protein interactions, trafficking and altered transcription. The C-terminal domain of NHE3 contains numerous putative phosphorylation sites for various kinases. In response to elevated intracellular cAMP, protein kinase A (PKA) phosphorylates NHE3 thereby inhibiting the activity of the transporter. The recruitment of PKA to the C-terminus of NHE3 involves a multiprotein complex with protein-protein interactions including NHE regulatory factors (NHERF1 and NHERF2) and a cytoskeleton-associated PKA anchoring protein (AKAP), ezrin. A similar scaffolding mechanism is responsible for glucocorticoid-stimulated NHE3 activity with a different mediating kinase, the serum and glucocorticoid inducible kinase (SGK1). SGK1 stimulates the activity of NHE3 by interacting directly with NHERF2, which acts as a bridge between the kinase and NHE3. Other studies demonstrate that a pool of NHE3 is localized to lipid rafts in the apical membrane and that NHE3 activity can be regulated by redistribution of the transporter between subcellular endosomal compartments and the plasma membrane. For example, epidermal and fibroblast growth factors (EGF and FGF, respectively) stimulate NHE3 activity by increasing the surface protein pool in a PI3-kinase-dependent manner [58, 124, 132]. In terms of transcriptional control, glucocorticoids also up-regulate NHE3 expression, but not that of NHE1, NHE2, or NHE4, consistent with the respective roles of the various isoforms in vectorial transport and housekeeping functions. In contrast, pro-inflammatory cytokines such as interferon γ and TNF α down-regulate NHE3 expression. Finally, recent studies suggest that, like CFTR (see Section 5.2.1), NHE3 may have a nontransport role in modulating the activity of other proteins due to its propensity to be engaged in protein-protein interactions.

5.1.1.4 Sodium Channel. The epithelial Na^+ channel, ENaC, also known as the amiloride-sensitive Na^+ channel is found on the apical membrane of specific epithelia; for example, it controls the absorption of sodium in the colon and reabsorption of Na^+ in the kidney (Figure 7, left cell). These selective ion channels are largely permeable to Na^+ ions but also conducts lithium and protons [83]. They are also found in the lungs and sweat glands and their presence on taste buds, can contribute as much as 20% of salt perception in humans. To maintain homeostasis, ENaC-mediated transport helps adjust the amount of sodium excreted in the feces, urine, and sweat. In the distal colon, ENaC is involved in the transepithelial Na^+ ion transport, which it accomplishes together with the basolateral Na^+/K^+ pump. The activity of ENaC is regulated by aldosterone, a mineralocorticoid secreted by the adrenal cortex in response to low blood pressure. The channel can be blocked by amiloride, which can be used as a diuretic to treat hypertension. The importance of ENaC for Na^+, K^+, and fluid homeostasis is emphasized by the fact that ENaC gain-of-function leads to extracellular volume expansion (Liddle's syndrome or pseudohyperaldosteronism). Similarly, a loss-of-function mutation lead to renal salt-wasting syndromes (pseudohypoaldosteronism type 1) accompanied with alterations in K^+ homeostasis.

ENaC is a heteromultimeric channel that is made up of three homologous α, β, and γ subunits that share 30% amino acid homology and belong to the degenerin/ENaC family [66]. Based on the primary structure, identified by expression cloning in *Xenopus* oocytes, two transmembrane helices were predicted. A large, glycosylated loop that separates the two transmembrane domains faces the extracellular side. The function of the extracellular domain is not known, but it has been suggested that the well-conserved extracellular domains might serve as receptors to control the activities of the channels. Hence, ENaC could very well be a ligand-gated channel. Further, it is predicted that the N- and C-terminus are facing the cytosolic side of the membrane [52].

ENaC interactions with CFTR have the most pathophysiological relevance in cystic fibrosis (CF). CFTR is a transmembrane channel responsible for Cl transport, and defects in this protein results in a cystic fibrosis phenotype (see Sections 5.2.1.2.2 and 8.4). CFTR has pleiotropic effects, affecting other ion transporters, including ENaC. In the colon, stimulation of CFTR leads to the activation of a basolateral K_VLQT_1-type K channel and inhibition of luminal ENaC. Ussing chamber studies done on epithelial biopsies from the colon of CF patients demonstrate that a complete defect of CFTR results in absence of Cl^- secretion and stimulated overabsorption of Na^+. This suggests an absence of ENaC down-regulation by CFTR in CF colon [37].

5.1.2 Potassium

Potassium transporters are an essential element of intestinal epithelial function. In addition to its role in the Na^+/K^+ pump, net transepithelial K^+ transport is key in helping the intestine balance fluid and electrolyte transport. Luminal K^+ concentrations generally increase from 6 to 9 mM in the jejunum to 13 mM in the ileum and 25 to 30 mM in the colon. Both K^+ secretion and absorption occurs down the length of the intestine with distinct segmental variations. In the small intestine K^+ transport is largely passive, whereas in the colon cellular transport processes contribute to transepithelial K^+ transport.

5.1.2.1 Potassium Absorption. Net K^+ transport from the lumen to the serosa in the small intestine, largely occurs by paracellular solvent drag, accompanying bulk fluid movement in response to a meal. This route of K^+ entry is affected in diarrhea and in severe cases can lead to hypokalemia. Net transepithelial K^+ transport in the colon is dependent on the luminal K^+ concentrations. Generally K^+ is secreted (see Section 5.1.2.2) when luminal $[K^+]$ is lower than 25 mEq/L. When concentrations rise above this, net absorption occurs, via primary active transport. The distal colonocytes and, especially in the rectum, possess K^+/H^+ ATPase pumps located on the luminal membrane (Figure 8). These pumps are P-type ATPases, and at least two colonic isoforms have been identified: a ouabain-sensitive isoform in the crypt cells and a ouabain-insensitive isoform in the surface cells. Both aldosterone and K^+ depetion up-regulate the ouabain-insensitive H^+/K^+-ATPase and stimulate K^+ absorption. These transporters belong to the same family as the K^+/H^+ ATPases found in gastric parietal cells.

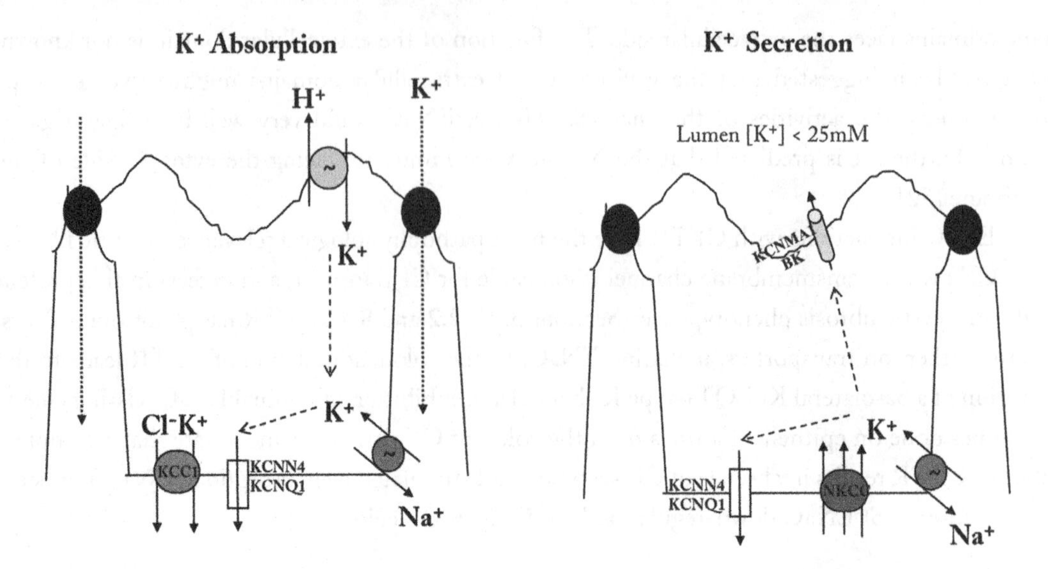

FIGURE 8: Potassium transport. In the small intestine, potassium is absorbed passively primarily through the paracellular pathway and by solvent drag. Potassium enters the cell basolaterally via the sodium pump and via NKCC; it leaves the cells through potassium channels (apical: KCNMA [BK]; basolateral: KCNN4, heteromer KCNE3/KCNQ1). In the colon, secretion can occur passively or through the cell. Net absorption takes place in the colon via apical H^+/K^+ ATPase pumps when the luminal concentrations are greater than 25 mEq/L.

5.1.2.2 Potassium Channels and Secretion. The increasing lumen negative potential difference with respect to the serosa, down the cephalocaudal axis (jejunum 5 mV, ileum 15–25 mV, and colon 30–50 mV) can lead to passive K^+ secretion in the small and large intestine. In addition, the colonic epithelia have apical membrane K^+ channels which contribute to net K^+ secretion when the luminal K^+ concentrations fall below 25 mEq/ L.

K^+ channels however play another critical role in the cell including intestinal epithelial cells. The activity of the Na^+/K^+ ATPase will eventually become rate limiting if there is no mechanism for the accumulated intracellular K^+ to exit the cell. Thus epithelial cells are endowed with K^+ channels, chiefly on the basolateral membrane that allow for the recirculation of K^+ to the serosal surface (Figure 8). The colonic epithelia, also possess apical K^+ channels that serve such a pump-leak function. K^+ exit results in a hyperpolarization of the membrane and serves further to stimulate pump-driven processes such as Na^+ absorption and Cl^- secretion.

K^+ channels are the largest group of ion channels in the human genome and a number of K^+ channels are shown to be present in the gastrointestinal tract. Different nomenclatures are used to name K^+ channels, causing considerable confusion. In this book, the nomenclature of the Human

Genome Organization is used. K^+ channels involved in association with secretion include the intermediate conductance KCNN4 channels and the KNCQ1 channels as determined by pharmacological inhibitor studies, biochemical localization, and knockout animals. Inhibition of these channels results in an attenuation of Cl^- secretion either via CFTR or, if extant, perhaps Ca^{2+}-activated Cl^- channels (see Section 5.2.1). Among the epithelial voltage-gated K^+ channels, KCNQ1's expression is shown to be the highest [105]. Evidence from KCNQ1 knockout mice studies and pharmacological data strongly imply that cAMP-activated Cl^- secretion largely depends on the activity of basolateral heteromeric KCNE3/KCNQ1 channels [47]. KCNN4 is shown to be not sensitive to Ca^{2+} per se, but it is activated by Ca^{2+} when it associates with calmodulin [47]. Matos JE et al. studied KCNN4, the big conductance Ca^{2+}-activated channels (KCNMA1) and KCNQ1 channels in wild type and knockout mice distal colon (Figure 8). They demonstrated that both KCNN4 and KCNQ1 channels are activated by cholinergic agonists and promote Cl^- secretion while KCNMA1 channels are not involved in this process [74]. Another possible K^+ exit pathway is KCC1 located on the basolateral membrane [99] (Figure 8, left cell).

KCNMA1 channels are apical membrane K^+ channels also referred to as the Big K (BK) channels, because of their large conductance (220 pS). They are voltage-gated, are Ca^{2+}-activated and are inhibited by verapamil, quinidine, or Ba^{2+}. They are primarily expressed in the surface cells and upper 20% of the crypt cells in the human colon [98] and are most likely associated with Na^+ absorption [6]. Like all K^+ channels, BK channel is a tetramer, with each monomer made up of an alpha and beta subunits. Each alpha subunit consists a transmembrane domain, a voltage sensing domain, a K^+ channel pore domain and a C terminal cytoplasmic domain containing Ca^{2+} binding sites. KCNMA1 channels are regulated by mineralocorticoids. No other apical K^+ channel has been reported to date in the human colon [98]. This is confirmed by the fact that in KCNMA1-knockout mice, colonic K^+ secretion is absent, suggesting that apical BK channels are the sole exit pathway into the lumen [103].

K^+ channels also play an important role in cell volume regulation and thereby help intestinal cells cope with the constant fluctuations in osmolarity. In response to cell swelling, K^+ channels are activated resulting in a regulatory volume decrease. A variety of K^+ channels are subject to modulation by membrane voltage, $[Ca^{2+}]_i$, pH, cell metabolism, and posttranslational modifications such as phosphorylation and sumoylation and are shown to affect cellular differentiation, aberrant growth in carcinogenesis and in apoptosis.

5.2 TRANSPORT OF ANIONS

The major anions transported across the intestinal epithelium are chloride, bicarbonate, SCFAs, phosphates, and sulfates. As in the case of cations, they utilize channels or transporters and exhibit heterogeneity both along the villus–crypt axis and along the cephalocaudal axis.

As noted in Section 5.1, for ease of reading, both the common functional names (for example, AE for anion exchangers) and the solute carrier family (SLC) nomenclature are provided.

5.2.1 Chloride

5.2.1.1 Chloride Absorption. Chloride and other anions are absorbed both via cellular and paracellular pathways. The lumen of the intestine is negative with respect to the serosal or blood side, and this transepithelial potential difference drives the movement of anions, including Cl^-, via the paracellular pathway in the jejunum. In the ileum and proximal colon, the absorption of chloride occurs via Cl^-/HCO_3^- exchangers coupled to Na^+–H^+ exchangers. In the distal colon, while Cl^-/HCO_3^- exchangers are present, the process is often independent of Na^+. Details on the Cl^-/HCO_3^- transporters are provided in Section 5.2.2.

There is still controversy on how Cl^- moves across the basolateral membrane during absorption [6]. In the proximal colon, apical absorption of Na^+ and Cl^- occurs through the concerted action of NHE3 and DRA. Basolateral Cl^- exit occurs either electroneutrally through AE2 (SLC4A2) or AE3b (SLC4A3). Inwardly rectifying Cl^- channels activated by hyperpolarization exist on the basolateral membrane of surface colonocytes, and the magnitude of Cl^- current in Ussing chamber experiments is equivalent to isolated colonocyte whole cell currents. There is some controversy on the localization of the Cl^- channel, CLC2 (see Section 5.2.1.4) in the intestinal epithelial cell. Locations ranging from the apical to the subapical region (below TJ) to the basolateral membrane have been implied (see Chapter 8). The role for CLC2 as the Cl^- channel in basolateral exit is indicated most strongly in that CLC2 and CFTR double knockout mice did not exhibit a further decrease in Cl^- secretion across the colonic epithelium.

5.2.1.2 Chloride Secretion. Fluid secretion in a variety of exocrine glands of the gastrointestinal tract is the result of the transepithelial movement of Cl^- from the blood side to the luminal compartment. Chloride moves through the cell by the orchestration of basolateral and apical membrane transporters, and the resulting electrical gradient drives the movement of Na^+ paracellularly; the ensuing osmotic gradient drives water movement into the lumen, resulting in net fluid secretion. In the healthy small and large intestine, a variety of factors including the electrochemical gradient of chloride, cell volume, autocrine, paracrine, immune, neuronal, and endocrine modulators as well as luminal substances maintain a basal rate of Cl^- secretion. This amounts to approximately 1.5 L of fluid per day and is necessary for the effective digestion and absorption of nutrients and the effective removal of waste products.

Although there are nuances in the regulation of Cl^- secretion in various regions of the gastrointestinal tract, the underlying cellular transport processes are rather ubiquitous (Figure 9). Thus, in

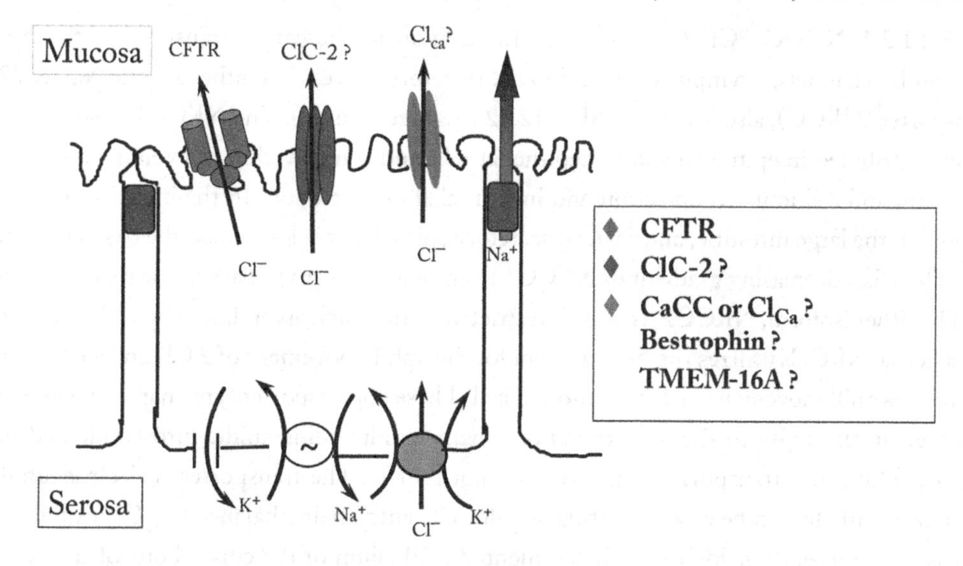

FIGURE 9: Chloride secretion. The Na^+/K^+ ATPase is the driving force for Cl^- secretion, but the process also requires a basolateral $1Na^+{:}1K^+{:}2Cl^-$ cotransporter, a K^+ channel, and an apical Cl^- channel. In the intestine, CFTR is the most likely candidate Cl^- channel. The role of other apical Cl^- channels such as CLC2 or Ca^{2+}-activated Cl^- channels need to be clearly established for the intestine.

most fluid-secreting epithelia, Cl^- secretion is electrogenic, and the driving force is the energy provided by the Na^+/K^+ ATPase located on the basolateral membrane. Located on the same membrane is an electroneutral cotransporter (NKCC1) that couples the movement of 1 Na^+:1 K^+:2 Cl^- and is responsible for Cl^- entry into the cell. The downhill entry of one Na^+ ion is sufficient to drive 1 K^+ and 2 Cl^- ions against their electrochemical equilibria into the cell. For effective functioning, the Na^+/K^+ pump has to be coupled to a "leak" mechanism and K^+ leaves the cell through K^+ channels located either on the basolateral membrane or in some tissues, such as segments of the colon, on the apical membrane. This results in an electronegative cell interior and in turn provides the driving force for Cl^- exit out of the cell via specific channels in the apical membrane of the cell. Thus, the careful coordination of the pump, NKCC cotransporter, and K^+ channels provide the driving force for Cl^- to enter the cell across the basolateral membrane, accumulate above its electrochemical equilibrium, and exit via Cl^- channels on the apical membrane. The flux of Cl^- across the apical membrane is electrically balanced by the exit of K^+ across the basolateral membrane. Blockage of the pump by inhibitors such as ouabain, of NKCC by loop diuretics such as bumetanide, of K^+ channels by barium, and of Cl^- channels by $CFTR_{inh}$-172 or GlyH-101 results in an inhibition of net Cl^- secretion.

5.2.1.2.1 Na^+-K^+-$2Cl^-$ Cotransport. In addition to the cation transporters, the Na^+/K^+ pump and K^+ channels, a symporter critical to the secretory process in epithelia is the $Na^+/K^+/2\ Cl^-$ cotransporter (NKCC), also known as SLC 12A2 (Figures 8 and 9). The NKCC1 isoform is ubiquitously distributed in epithelial and nonepithelial cells and is involved in cell volume homeostasis, cell growth, and cell ionic composition and in epithelial ion secretion. In the intestine, it is highly expressed in the large intestine, and as in all secretory epithelia, it is located on the basolateral membrane. There is a decreasing gradient of NKCC1 from cells at the crypt base to the surface epithelial cells. The other isoform, NKCC2, is largely restricted to the thick ascending loop of Henle and the macula densa. NKCCs utilizes the Na^+ gradient for the uphill movement of 2 Cl^- and a K^+ ion along with the downhill movement of 1 Na^+ into the cell. These cotransporters are characterized by their rank order of sensitivity to the loop diuretics benzmetanide>bumetanide>furosemide and halide selectivity. Thus, they transport Cl^- and Br^- but not F or I^-. The transporter is electroneutral and in secretory epithelia can be a rate-limiting step for Cl^- entry, being balanced by Cl^- exit via apical channels. More recently, evidence in the pigmented epithelium of the ciliary body of the eye indicates that like the Na^+–glucose cotransporter, NKCC1 can transport water (1 Na^+:1 K^+:2 Cl^-:570 molecules of water).

The NKCC1 protein is a 130- to 140-kDa, 12 transmembrane glycosylated protein with a long, intracellular N-terminal domain that is not very well conserved among species, and a long, intracellular C-terminal domain that is highly conserved. NKCC1 is subject to regulation by a variety of signaling pathways, and this could involve one or more of the following: increasing trafficking of NKCC1 to and from submembrane endosomal compartments, altering the activity of the transporter (via phosphorylation), or long-term gene regulation and examples are provided below.

As an example of trafficking, forskolin largely stimulates activity without altering membrane abundance of NKCC1, whereas cholinergic stimulation increases recruitment of transporters to the basolateral membrane. The activity of the cotransporter is indirectly regulated by changes in $[Cl^-]_i$, in cell volume, by nutrients and modulators involving a number of cytoskeletal interactions, phosphatases, and kinases. Thus, a decrease in $[Cl^-]_i$ or cell volume after anion secretion increases NKCC phosphorylation and its activity. The NKCC1 protein possesses consensus phosphorylation sites for a number of kinases, including PKA and PKC, but there is little evidence for direct phosphorylation by these kinases. It is also unclear if tyrosine kinases play a direct role in NKCC phosphorylation in the intestine. Phosphorylation by a number of activators including cAMP, cell shrinkage, and fluoride are at similar sites suggesting a common kinase. The kinase is inhibited by cell swelling, NEM, and staurosporine. A unique proline–alanine-rich STE20-related kinase appears to be the enzyme that regulates the bulk of the activity. In the T84 colonic cell line, NKCC stimulation by cAMP, appears to involve phosphorylation of myosin light chain kinase. Not all kinases are stimulatory, and PKC and the Ser–Thr WNK4 kinase inhibit NKCC, and small interfering RNAs directed against novel PKCs reverse the effect. PKC, specifically PKC ε, may also be involved in trafficking by phosphorylating cross-linking proteins and disassembling cortical cytoskeleton. Furthermore, the cytoskeletal disrupting agent, cytochalasin D, prevents inhibition of NKCC by the PKC activators, phorbol esters. Other modulators regulate gene expression of *NKCC1* as seen with 24 hr. exposure of colonic epithelial cells to EGF, resulting in a sustained increase in Cl^- secretion. Butyrate, proinflammatory cytokines, and γ interferon decrease NKCC activity, whereas TNFα, IL6, and IL1 increase its activity. In contrast, inflammatory mediators, H_2O_2, and hypoxia-inducible factors (HIF1) down-regulate NKCC to prevent Cl^- loss during hypoxia; these agents also down-regulate basolateral membrane K^+ channels, the net result being a decrease in the driving force for Cl^- exit.

5.2.1.2.2 CFTR Chloride Channel. In secretory epithelia, the major mechanism for Cl^- exit via the apical membrane is through Cl^- channels, and at least three separate classes of Cl^- channels with distinct electrophysiological characteristics have been identified (Figure 9). The most prominent, both in terms of expression and activity, is the CFTR, a linear, low-conductance channel. CFTR is strongly expressed in the apical membrane of colonic and small intestinal crypts and to a more modest extent in similar regions of villar and surface cells. Functionally, it is a linear, low-conductance (4–12 pS) channel with a halide ion selectivity of $Br^- > Cl^- > I^- > F^-$ and is activated by cAMP. It is a protein with pleiotropic effects, and in addition to transporting halides, it transports HCO_3^- and ATP and can influence the expression and activity of a variety of other proteins in both apical and basolateral membranes. For example, the presence of apical membrane CFTR can influence apical Na^+ channels, outwardly rectifying Cl^- channels, AEs as well as transcriptionally and functionally increasing basolaterally located NKCC and Na^+-HCO_3^- (NBC) proteins. Furthermore, CFTR and other transporters can share scaffolding proteins and thereby modulate function, e.g., the C-terminal of CFTR interacts with NHERF, a protein that also binds to NHE3.

Structurally, CFTR belongs to the super family of ATP-binding cassette proteins, of which the multidrug-resistant *p*-glycoprotein is a member. The *CFTR* gene was the first to be identified by linkage analysis and positional cloning and spans 24 exons. This codes for a large (250-kDa) glycoprotein, which undergoes extensive posttranslational modifications in the Golgi complex and endosomes before being targeted to the apical membrane. In the cell, CFTR channels reside in both a recycling endosomal pool and in the apical plasma membrane (Figure 10). The protein has five subunits and topologically can be roughly divided into halves separated by a large cytoplasmic regulatory or R domain; each half has six transmembrane helices (TM1–6 and TM7–12), forming the ion pore, followed by a nucleotide binding domain (NBD1 and NBD2) (Figure 10, inset). Although there is considerable sequence homology between the two membrane domains and between NBD1 and NBD2, they are not identical. The R domain has a number of kinase consensus

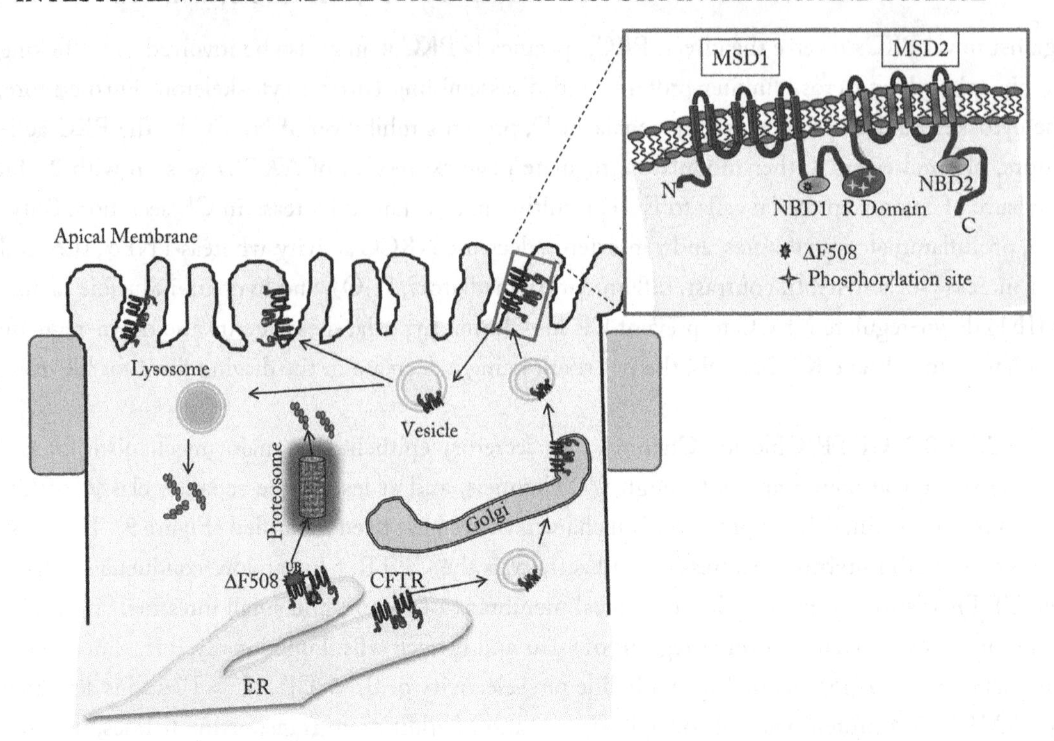

FIGURE 10: CFTR. CFTR is an apical membrane Cl^- channel. Inset: The protein has two 6-helical transmembrane domains (MSD1 and MSD2) which form a pore, 2 nucleotide binding domains (NBD1 and NBD2) that bind ATP and regulate channel opening and closing, and a central regulatory domain that must be phosphorylated by PKA for ATP binding to NBD. Main Figure: The protein is normally trafficked to the apical membrane via the Golgi and endosomal pathway. cAMP increases activity of the channel and insertion of channels into the membrane. A mutation in CFTR (δF508) results in a misfolding of the protein, leading to ubiquitinylation and thereby marking the protein for degradation via the proteasomal and lysosome pathway. The net result is very few, if any, δF508 CFTR proteins get translocated to the apical membrane and those that do show altered activity.

sequences, especially for PKA. The cytoplasmic N and C terminal domains are the sites of regulatory influence upon CFTR but also of CFTR's influence on other proteins via protein/protein interactions, anchoring proteins, and consequent phosphorylation events.

Structure–function studies, including the use of specific mutations have revealed that gating and activation of the channel occurs through a well-ordered sequence of events, initiated by the phosphorylation of the R domain by cAMP-dependent PKA. This leads to a conformational change in NBD1 and allows for it to bind and hydrolyze ATP, resulting in a flickering opening of the channel. Further, phosphorylation of the R domain changes the conformation of NBD2, allowing it to bind to ATP. This leads to a stable opening of the channel. Hydrolysis of ATP at NBD2 returns the channel to a flickering open state and eventually to the closure of the channel. Thus, both phosphorylation by PKA and ATP binding are needed for CFTR to operate. Regulation of the channel occurs via altering the open probability of the channel (opening/closing) as well as by altering the number of CFTR molecules in the membrane by promoting trafficking between the recycling endosomal pool and the apical membrane. Thus, increases in cAMP activate PKA, which increases Cl^- secretion by simultaneously activating CFTR as well as increasing channel number by recruiting CFTR-bearing endosomes to the apical membrane [63, 92, 94].

Defects in the *CFTR* gene lead to the autosomal recessive disease CF, which affects exocrine tissues including the intestine. The major causes of mortality are respiratory disease and pancreatic insufficiency arising out of the basic CF defect (see Section 8.4). Over 1500 mutations of CFTR have been identified, and their effects on CFTR function been broadly grouped into six classes. However, it must be noted that close to 75% of CF patients carry the delta F508 mutation, i.e., the deletion of a single phenylalanine residue. The classes based on molecular phenotype are as follows: (I) no protein as a result of nonsense or frameshift mutations or abnormal mRNA splicing; (II) less than wild-type protein in apical membrane due to intracellular quality control during protein processing; the delta F508 deletion results in a misfolding of the protein and marks it for proteasomal degradation in the lysosome (Figure 10); (III) defective regulation of CFTR at the apical membrane, generally resulting from nucleotide substitution; (IV) defective anion transport through CFTR in the apical membrane; (V) reduced synthesis of normal CFTR, resulting in decreased protein amounts; (VI) reduced residence time in the apical membrane of mature protein generally arising from truncations in the C terminus.

Approximately 15% of CF adults present with distal intestinal obstructive syndrome, and a common pathology in the CF infant is meconium ileus. The defect in the CF mouse is manifested largely as meconium ileus in the intestine (see Section 8.4). In the intestine, abnormal secretion of mucopolysaccharides and their inadequate hydration are a consequence of the CF disease. In any manifestation of the disease, it remains to be determined if the pathology is due to changes secondary to decreased fluid secretion or result from more direct effects of CFTR on other protein function. For example, CF mice have goblet cell hyperplasia, increased inflammation, increased crypt cell proliferation, Paneth cell abnormalities, with decreased mCLCA3 (a Ca^{2+}-activated Cl channel; see Section 5.2.1.2.3) expression/cell. The severity of the disease appears to be inversely proportional to the expression of mCLCA3 (its human orthologue is hClCA1). Mucus production/secretion is exaggerated, with engorged crypts and occluding mucus in CF mice, and survival rates are poor. However, correction of the mCLCA3 expression results in a decrease in mucus accumulation and improved survival, but with continued hyperplasia and aberrant responses to the adenylate cyclase activator, forskolin. Thus, genetic modifiers can help relieve the disease symptoms but not alter the basic CFTR defect. Recent therapeutic strategies to treat CF or diarrhea is to use combinatorial chemistry to generate series of compounds that can specifically inhibit or stimulate CFTR [118] (see Section 8.4).

5.2.1.2.3 ClC Chloride Channels. A distinct and ubiquitous family of Cl^- channels are the ClC family. This family has 9 isoforms, of which CLC-0, CLC-1, CLC-2, CLC-4, and CLC-5 have been demonstrated to form functional Cl^- channels; while many of them are either tissue-specific or localized to subcellular compartments, the CLC2 isoform is a widely distributed plasma membrane protein [56]. The *CLC2* gene codes for a ~97-kDa protein that functions as an inwardly rectifying Cl^- channel. The channel shows ion selectivity and is inhibited by Cd^{2+} and a peptide toxin GaTx2.

The function of ClC2 has been linked to regulation of cell volume, intracellular pH, and intracellular Cl^-. However, the distribution of this protein in the intestine and its role in Cl^- secretion are still the subject of debate (Figure 10). The protein has been localized both to the apical membrane of T-84 colonic cell lines, a model of the Cl^- secretory epithelium, and to the lateral membranes just below the TJs in Caco-2 (apical aspect of the TJ) cells, another model of colonic epithelia. In contrast, in the mouse colon, it appears to be localized to the basolateral membrane since ClC2 cannot rescue Cl^- secretion in CF–/– mice, and furthermore, double knockout ClC2–/–+CFTR–/– mice shows an improved intestinal secretion (phenotype) as compared with CFTR–/– mice. Both these features have been interpreted to indicate a basolateral location of ClC2 in the mouse colon.

Much of the debate centers around the contention that CLC2, and not CFTR, is the specific target for the novel prostanoid laxative, lubiprostone. Some investigators say that ClC2 is a unique target site, not regulated by cAMP, and therefore lubiprostone has the potential for use in CF patients as a potential activator of a salvage pathway and therefore could be of therapeutic benefit in the treatment of CF. These authors localize CLC2 to either the apical membrane or to regions that are just below the TJs on the basolateral membrane. However, other investigators, including the authors of this book [4], have demonstrated that lubiprostone activates CFTR in a classical cAMP-dependent manner; furthermore, some authors have convincingly demonstrated that lubiprostone cannot stimulate Cl^- secretion in intestinal epithelia devoid of CFTR (cell lines and the CF–/– mouse) (10). Clearly, much more work needs to be done to elucidate the interplay between ClC2 and CFTR in the functioning of the human intestine in health and disease [4, 19].

5.2.1.2.4 Ca^{2+}-Activated Cl^- Channels. Another class of Cl^- channels, termed the calcium-activated Cl^- channels (CaCC, Figure 10), has been long recognized based on their functional properties and regulation: these channels are outwardly rectifying, activated by Ca^{2+}, in many cases involving CAM kinase II, and inhibited by niflumic acid. While functional studies have firmly established their presence in glandular organs such as the salivary gland, lacrimal gland, and pancreas and in respiratory epithelia, the definite demonstration of their activity in the intestine is much more robust in intestinal epithelial cell lines, such as the T84 cells, than in native intestinal epithelia. Nonepithelial tissues such as neurons and muscle cells, including intestinal smooth muscle cells, also exhibit Ca^{2+}-activated Cl^- transport.

The molecular characterization of the calcium-activated Cl^- channels has eluded scientists for a long time, with a number of potential candidates being postulated and often with differing nomenclature (e.g., CLCA, bestrophins, generic CaCC). The isoform ClC3 of the ClC family can be activated by Ca^{2+}-CAM kinase II but has properties distinct from the outwardly rectifying current identified in many tissues. The CLCA family of proteins detected both in intestinal and airway epithelia was considered to be the Ca^{2+}-activated Cl^- channel, but there is some uncertainty whether the members are truly membrane proteins, as they have been found in cytosolic fractions. Nevertheless, as discussed in the previous section, there is evidence that mouse CLCA (mCLCA3 orthologue of human CLCA1) may play a role in goblet cell function and their expression appears to be down-regulated in the absence of CFTR. Restoration of CLCA to these mice ameliorates the severity of the disease but does not restore CFTR function. The bestrophins, which are the product of the vitelliform macular dystrophy (*VMD*) gene, were proposed to be the universal epithelial Ca^{2+}-activated Cl^- channel. However, this oligomeric channel, which is Cl^--sensitive, lacks the characteristic outwardly rectifying kinetic properties of the Ca^{2+}-activated Cl^- current and is not activated by all Ca^{2+}-agonists but is activated by increases in cell volume. Therefore, these channels may be more volume-sensitive Cl^- channels associated with pigment epithelia. Using distinct approaches, recently, three separate groups of investigators have definitively identified TMEM-16A (ANO1) as the epithelial calcium-activated Cl^- channel (CaCC). It belongs to a unique family of 10 proteins, the anoctamins, so named because they have eight putative transmembrane segments. Not all the members of this family are believed to exhibit Ca-activated channel behavior. Mice lacking TMEM-16A do not express any Ca^{2+}-activated Cl^- secretion. Evidence suggests that these are the channels per se, but the mechanisms by which they are regulated (via auxiliary proteins, they act as regulators per se, etc.) remain to be determined [29].

What is the evidence for Ca^{2+}-activated Cl^- channels in intestinal tissues? As mentioned above, the demonstration is much more definitive in cell lines than in native tissues. In native tissues, rotavirus infection, specifically the rotavirus toxin NSP4, has long been associated with Ca^{2+}-dependent Cl^- secretion in the colon of the young mammal, including in 7- to 14-day-old CF−/− mice [77]. However, more recent evidence by Sepulveda et al. [30] failed to consistently demonstrate that carbachol induces a Ca^{2+}-activated Cl^- current in 8- to 14-day-old or adult CFTR−/− mice; they do exhibit K^+ secretion. In addition, the carbachol-stimulated Cl^- current in wild-type mouse colon was not sensitive to niflumic acid, and while apical addition of ATP or UTP caused an increase in Cl^- secretion in respiratory epithelia, consistent with a Ca^{2+}-activated Cl^- current, in colonic epithelia, only a transient K^+ secretion was observed [29]. Using combinatorial chemistry, a series of compounds capable of activating TMEM-16A without increasing $[Ca^{2+}]_i$ have been identified. These compounds increase CaCC activity in the salivary glands, IL4-treated bronchial cells, submucous respiratory glands, and intestinal smooth muscles. In the intestinal muscle, TMEM-16A has been specifically localized to the interstitial cells of Cajal, the pacemaker cells, and the protein has a functional role in smooth muscle contraction. Interestingly, intestinal epithelial cells express message for TMEM-16A, but there is no evidence for its function on the apical membrane. Although, CFTR–/– mice do not exhibit carbachol-stimulated apical Cl^- secretion, they do exhibit carbachol-induced changes in cell volume, which has been interpreted to be due to a basolateral conductance. Taken together with other studies suggesting the presence of Cl^- conductance on the basolateral membrane, it is possible that TMEM-16A functions as a basolateral channel in the intestine and an apical channel in other tissues [29].

Thus, a growing body of evidence suggests that in native intestinal epithelia, the majority of chloride secretion occurs through CFTR, and the role of CaCC is perhaps restricted to other regions of the intestine, especially the smooth muscle.

5.2.2 Bicarbonate

Bicarbonate is the major anion in the stool in many diarrheas and plays a critical role in fluid balance in the intestine. It contributes to the alkalinization of intestinal chyme in the duodenum and is also secreted along the length of the cephalocaudal axis, including the ileum and the colon. A number of mechanisms contribute to intracellular bicarbonate; it can be generated by metabolism, by diffusion of CO_2 into the cell and subsequent action of carbonic anhydrase, and/or by basolateral anion transporters, such as the Na^+:HCO_3^- cotransporter (*SLC4A4 and SLC4A7*). The bulk of bicarbonate secretion occurs via electroneutral Cl^-/HCO_3^- exchangers, on the apical membrane of the enterocyte and colonocyte (Figure 7). In tissues such as the pancreas and cholangiocytes, Cl^- exiting the cell via CFTR is thought to be recycled back into the cell via Cl^-/HCO_3^- exchangers, resulting in a net alkaline secretion. In the colon, HCO_3^-/SCFA$^-$ exchange is an important transporter and contributes to HCO_3^- secretion. Some secretion can be attributed to electrogenic bicarbonate secretion that chiefly occurs through channels, generally CFTR.

Unlike the "absorptive NHE isoforms" (e.g., NHE3), which tend to be concentrated in the villus regions, AEs are present on the apical membranes of both crypt and surface cells. The AEs (Cl^-/HCO_3^- exchangers) were first identified in the red blood cell (AE1), and indeed the abundance of this protein in the red cell membrane (also known as Band 3) allowed for its ready biochemical and molecular characterization. The AE isoforms in other cell types, including the intestine, are much less abundant, and the identification and characterization of the epithelial isoforms AE2, AE3, and AE4 was much slower. The AEs are now classified as the SLC4 bicarbonate transporters, and their role and regulation in both apical and basolateral membrane intestinal ion transport are still being elucidated. There is some indication that a separate HCO_3^- conductive pathway may mediate bicarbonate secretion into the duodenum.

A group of transporters gaining increasing prominence in epithelial anion transport are members of a distinct multifunctional anion exchange family SLC26 because of their wide distribution and ability to transport a variety of anions. In addition to Cl^- and HCO_3^-, the SLC26 proteins can transport, with differing affinities, sulfate, formate, oxalate, hydroxyl ions, and other anions. Of special interest are SLC26A3 and SLC26A6, which show an interesting differential expression down the length of the cephalocaudal axis. *SLC26A3* was first identified as the gene *DRA*, and mutations in this gene causes congenital chloride diarrhea, a disease presenting with severe volume depletion and metabolic alkalosis. SLC26A3 transports >2 Cl^-:1 HCO_3^- ion and is abundantly expressed on the apical membranes of colonocytes but not of enterocytes (Figure 6). In marked contrast, SLC26A6 is abundantly expressed in the apical membrane of villus enterocytes and less so in the colonocytes (Figure 5). This exchanger also referred to as the putative anion transporter (PAT-1) transports > 2 HCO_3^-:Cl^- ion. Absence of DRA (SLC26A3) results in severe diarrhea, increased colonic proliferation and up-regulation of other transporters in ion absorption in the colon, underscoring the importance of this transporter in neutral NaCl absorption. In contrast, SLC26A6-/- animals do not show a pathological phenotype, suggesting that other compensatory mechanisms are operative [107, 120]. As in the case of NHEs these proteins undergo both short-term regulation and long-term regulation and considerable focus has been given to the regulation of DRA (SLC26A3). The regulation of DRA and PAT-1 is functionally linked with that of NHEs and in certain cell types, protein-protein interaction through PDZ adaptor proteins have been reported [65]. Thus, DRA and NHE3 are both inhibited by cAMP and Ca^{2+}, a function associated with clathrin-mediated endocytosis [81]. DRA has been localized to lipid rafts, and its association and activity is increased by the proabsorptive neuropeptide Y (NPY) [96]. In terms of long-term regulation glucocorticoids thus increase the expression of DRA and PAT-1, much like it does NHE3; butyrate and commensal bacteria, such as Lactobacillus, up-regulate the expression of these anion exchangers. As in the case of NHEs, proinflammatory cytokines reduce the expression of DRA and PAT-1, and this is in keeping with the reports that in animal models of inflammation as well as in tissues from IBD patients, there is a reduction in NaCl absorption. The coordinated regulation of Na^+ and Cl^- transporters and therefore of NaCl absorption is teleologically sound [71].

Basolateral HCO_3^- uptake is accomplished by the NBC1 protein also known as SLC4A4 or its homologues A7 and A5—all expressed to varying degrees down the gastrointestinal tract and exocrine glands. The relevance of different isoforms remains to be delineated. NBC1 (SLC4A4) is critical for HCO_3^- reabsorption, as its deletion results in systemic acidosis. NBC1 can be regulated by cAMP (both increase and decrease, with the latter dependent on NHERF1) and by IRBIT, the inositol 1,4,5-trisphosphate (IP3) R Ca^{2+}-release channel-binding protein, which promotes NBC1 activity. Exocytosis is an important step in regulating NBC1 activity. In conditions of chronic alkalosis and acidosis, a concomitant increase and decrease of NBC1 was noted in various organs, including the gastrointestinal tract; this underscores the role of NBCs in regulating systemic pH.

Like other transporters, HCO_3^- secretion is regulated by a number of signal transduction cascades, including those involving cyclic GMP (cGMP). The functional linking of recycling of Cl^- transport through CFTR with Cl^-/HCO_3^- transport via exchangers is efficient, but it is interesting that CFTR appears to regulate the expression of anion exchange proteins as well. It is intriguing that there are a plethora of anion transporters, and the physiological relevance of their distribution along the villus–crypt axis, apical vs. basolateral membranes, and also along the cephalocaudal axis remains to be fully elucidated.

5.2.3 Short-Chain Fatty Acid

A major source of metabolic fuel in the colon are the SCFAs, and the 2–4 carbon SCFA (e.g., acetate, propionate, butyrate) are the major anions (60–150 mmol/kg) in the colon (Figure 11). They are generated by the action of the bacterial microflora on undigested and poorly absorbed carbohydrates, and the luminal concentrations generated set up a large concentration gradient across the colonic epithelium. In addition to providing energy, SCFA regulate colonocyte growth and differentiation and are used in therapeutic strategies for addressing colonic inflammatory diseases (e.g., bypass colitis). Colonic SCFA absorption parallels that of Na^+ and occurs by a number of mechanisms. SCFAs are weak electrolytes and can be protonated or ionized. The protonated species can diffuse across the apical membrane of the colonocyte. However, at the luminal colonic pH of 6.4–7.5, SCFAs are 95%–99% ionized and need specific carriers to be transported. Transport of SCFAs is linked to Na^+ transport mechanisms, and SCFAs are known to increase the expression of the NHE3 on the apical membrane. The net result is Na^+-SCFA and fluid absorption. Current views on SCFA absorption suggest that apical NHEs may create an acidic pH microclimate at the surface, which could promote the diffusion of SCFA-H into the cell. In addition, there are specific members of the SLC families, SLC16 and SLC5, involved in electroneutral, carrier-mediated transport of SCFA. These transporters are also referred to as monocarboxylate transporters (MCTs), SLC16A1 (MCT1), and transport 1 H:1 SCFA, and require an ancillary protein for their function [40]. In addition, two members of the SLC5 family, the high-affinity (SLC5A8) and low-affinity (SLC5A12) Na^+-dependent MCTs, have been implicated as having a role. Finally, other exchangers, Cl^-/butyrate exchangers and $SCFA/HCO_3^-$ exchangers, have been functionally related to Na^+/H^+ exchange. The molecular identity of these activities and how they account for SCFA promotion of electroneutral Na^+ and Cl^- absorption remains to be established [11, 31].

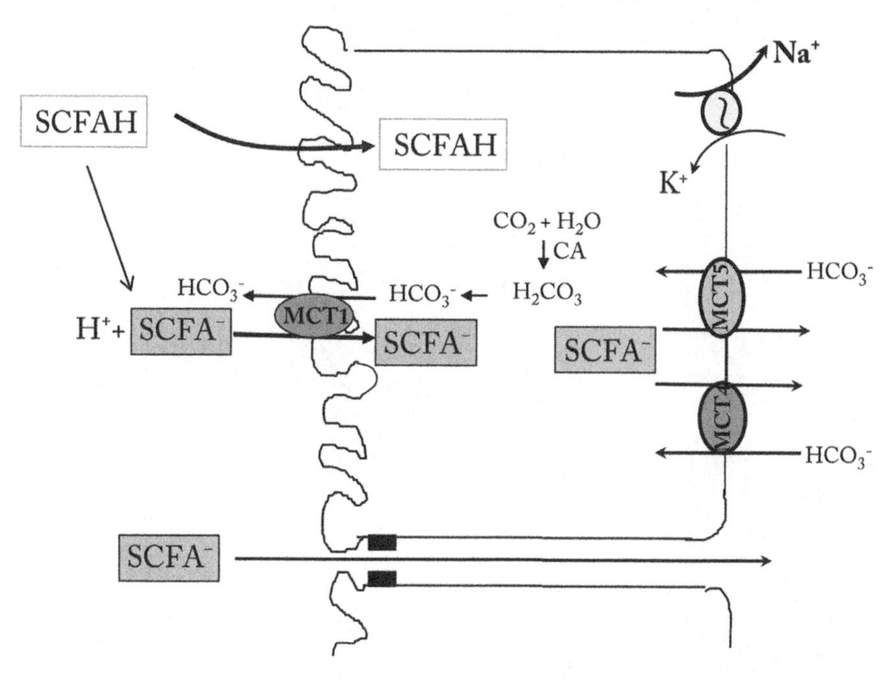

FIGURE 11: Short-chain fatty acid (SCFA) transport. The protonated SCFA (SCFAH) diffuse across the apical membrane of colonocytes. Most SCFA are ionized, and members of the SLC families, monocarboxylate transporters (MCTs), are involved in the electroneutral, carrier mediated transport of SCFA. $SCFA^-$ can also cross the epithelium through the paracellular pathway.

• • • •

CHAPTER 6

Water Transport

A major property of the intestine is the transepithelial movement of water at rest and in response to a meal; this involves both intestinal secretions as well as absorption accompanying nutrients in the small intestine and reabsorption in the colon. The movement of water is linked to the movement of solutes, and until recently, it was thought that this occurred strictly in response to osmotic gradients. Over the past two decades, a clearer picture of how the movement of this essential molecule occurs and is "regulated" has emerged. For over 40 years, the standing gradient hypothesis was the explanation of the transepithelial movement of water across the intestine. This hypothesis states that a small increase (2–3 mOsm) in the osmolarity across the apical membrane is sufficient to drive water into the cell; a subsequent similar increase across the intercellular and subepithelial spaces, in combination with hydrostatic gradients and relatively high water permeabilities, causes movement of water across the epithelium, both through and around the cells [24]. While an elegant explanation, the hypothesis is not borne out with experimental evidence and does not fully account for some fundamental physiological properties. Thus, electrophysiological measurements in lateral intercellular compartments with ion-selective microelectrodes failed to reveal any hyperosmotic compartments [133]. Similarly, while the geometry of lateral spaces and epithelial architecture may prevail as important diffusive barriers in some highly specialized intestinal structures such as the flounder intestine, they may have a less significant role in the mammalian intestine.

Both in absorption and secretion, epithelia can either transport water uphill, against a transepithelial osmotic gradient, or exhibit "isotonic" water transport, i.e., in the absence of a lumen→serosal osmotic gradient. In the intestine, the former is best exemplified in response to a meal. The digestion of proteins and polysaccharides both by cavital and surface hydrolases can increase intestinal osmolarity by as much as 250 mOsm as compared to the plasma. While large increases in luminal osmolarity can shut down water transport completely, most epithelia continue to transport water uphill in the face of modest increases in osmolarity, albeit with a reduction in the rate of water transport. The lumen→plasma osmolarity gradient, at which there is a cessation of water transport, varies with the tissue and in the mammalian intestine is 150–250 mOsm, whereas it is 30–60 mOsm in the kidney and 80 mOsm in the gall bladder [133]. Similarly, in secretory tissues such as the salivary gland, transport against 60-mOsm osmotic gradient or 300 cm H_2O hydrostatic pressure gradients from plasma to lumen can occur.

6.1 ROUTES OF WATER TRANSPORT

Calculations estimate that as many as 175 molecules of water are transported per ion or molecule of solute. So how is the transport of water achieved in the intestine and what are the mechanism(s) coupling it to the movement of solutes? There is now good evidence that water transport across epithelial cell membranes occurs by one of two processes and employs one or more of at least four routes. The two processes are osmosis, which is passive, governed by differences in the chemical potential of water and hydrostatic pressure and cotransport where the movement of water is coupled to energized by the movement of solutes and where a coupling ratio of solute/water molecules exist. The four routes include diffusion through the lipid bilayer, water channels, uniporters, and cotransporters. Since the intestinal epithelium is both absorptive and secretory, it is not surprising that different combinations of these routes are employed in the function of fluid absorption vs. fluid secretion (Figure 12).

Initially, transport of water was thought to occur only by dissolution of water through the lipid bilayer and that the composition of the bilayer dictated the relative permeability of the membrane to water. Nevertheless, some epithelial cells have relatively high water permeability that could not be explained solely by osmolarity changes. In the early 1950s, "water pores" were postulated as the additional mechanism for water transport across membranes. However, it was not until the late 1980s, when two groups of scientists identifying a membrane protein involved in water transport in erythrocytes [8, 20], led to the remarkable discovery of the family of water transport proteins, the AQPs.

6.2 AQUAPORINS

6.2.1 Aquaporin Structure and Function

Currently, there are 13 identified mammalian AQPs, constituting a family of small (≈28 kDa), hydrophobic, membrane proteins. AQPs are ubiquitous, being found throughout the bacterial, plant, and animal kingdoms. The distribution of AQPs in mammals is tissue-specific and they are subject to different forms of regulation. While initially identified as a water channel, it soon became clear that some AQPs also transport small hydrophilic molecules such as urea and glycerol. The AQPs function as a tetramer made up of four identical monomers forming a central pore [117]. Each monomer has six transmembrane helices and two shorter helices extending into the plane of the membrane. Transport is strictly governed by osmosis and a hydrostatic gradient within the pore, although gating of AQPs is known to occur under special circumstances especially in plants. The

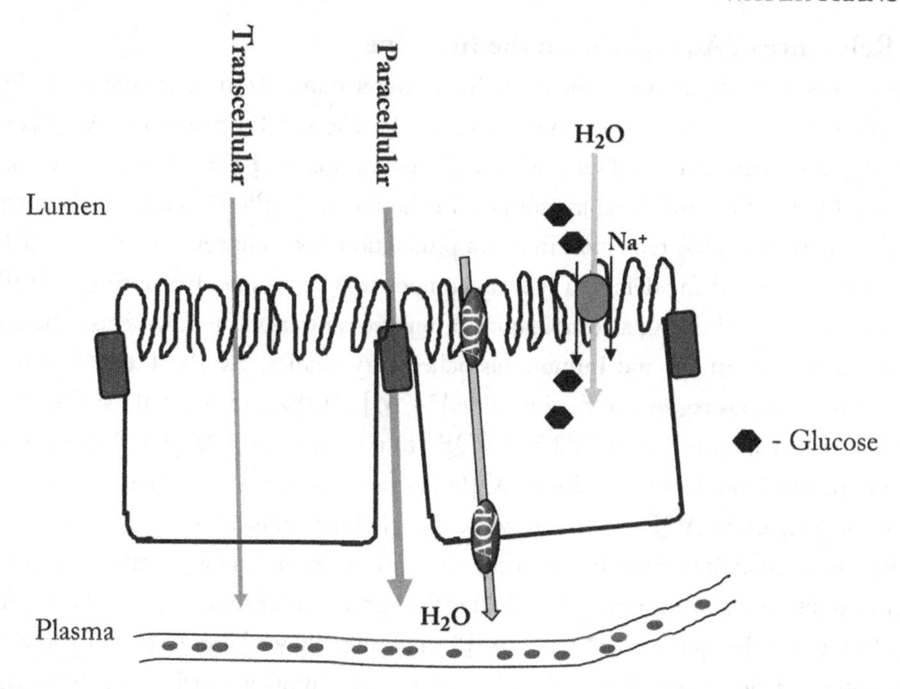

FIGURE 12: Water transport. Water transport across epithelial cell membranes employ one or more of at least four routes: diffusion through the lipid bilayer, water channels (aquaporins, AQPs), uniporters, and cotransporters (e.g., SGLT, NKCC).

permeation pathway allows for the single-file diffusion of water and is impermeable to charged species. AQPs have two filters that prevent the movement of ions; the first highly conserved, "NPA" (asparagines–proline–alanine) filters out most ions and some H^+ and the second an "aromatic arginine," which effectively filters out H^+. The second filter is variable and determines if hydrophilic molecules can permeate. Thus, AQP1, AQP2, AQP4, AQP5, and AQP8 are primarily water-selective, whereas AQP3, AQP7, and AQP9 also transport glycerol and other small solutes and are termed "aquaglyceroporins" [119]. Oxygen and carbon dioxide can also be transported through AQPs. While AQPs clearly increase water permeability across a tissue, their role in overall water homeostasis is unclear at best. Thus, individuals who lack the constitutively expressed AQP1−/− only exhibit a phenotype under conditions of severe dehydration, and systematic deletion of the four *AQP* genes in the nematode animal model, *Caenorhabditis elegans*, has no phenotypic effect. The compensatory mechanisms presumably reside in other mechanisms of water transport or in the other AQP isoforms and remain to be determined. However, impaired function of AQPs have been implicated in nephrogenic diabetes insipidus and in congenital cataracts.

6.2.2 Relevance of Aquaporins in the Intestine

Based on the tissue distribution and the centrality of water transport in gastrointestinal physiology, considerable attention has been given to studying the function and regulation of AQP family proteins in the gastrointestinal tract. Thus, salivary glands express AQP5 in the apical membrane and AQP1 and AQP4 in the basolateral membrane; furthermore, AQP5–/– mice exhibit a hypertonic saliva with a more than 60% reduction in saliva production [69]. However, neither AQP1–/– nor AQP4–/– mice exhibited any defect in salivary composition or volume. Interestingly, AQP5 is not found in the intestine. Utilizing a combination of reverse transcription–polymerase chain reaction (RT-PCR), immunoblotting, and immunohistochemistry studies, AQP1–4 and AQP6–10 have been localized to various regions of the intestine [54, 61]. AQP7 and AQP10 have been localized to regions of the small intestine, AQP2 and AQP4 to the colon, and AQP 3, AQP6, AQP8, and AQP9 to both small and large intestines. AQP4 has been localized to the basolateral membrane of surface colonocytes. In AQP4–/– mice, while theophylline-induced fluid secretion was not impaired, there was a modest increase in fecal water content, suggesting a lower transepithelial osmotic water permeability and a role for AQP4 in fluid reabsorption rather than secretion [121]. Although AQP8 is located on the apical membrane, AQP8-null mice showed a mild phenotype with active fluid absorption, cholera toxin-driven fluid secretion, and osmotically driven water transport, remaining essentially unimpaired in midjejunal or colonic loop studies [130]. More recently, Ikarashi et al. [53, 54] reported that magnesium sulfate administration up-regulated colonic luminal AQP3 protein expression in rats, possibly through the PKA-CREB pathway. Fecal water content increased in response to $MgSO_4$, reaching severe diarrhea proportions 4–8 hours after administration. Thus, although considerable attention has been given to exploring the role of AQPs in the intestine, the AQP knockout studies have yet to pinpoint a specific functional intestinal water channel, if any.

The lack of a clear role for AQP in the intestine may rest with the nature of the milieu and function of the intestine. Thus, since transport of water through AQPs is osmotically driven, the presence of water channels in the apical membrane when the gut faces a hyperosmotic lumen→cell gradient as seen during digestion would only serve to promote water loss rather than water absorption. In contrast to the intestine, the renal proximal tubule, a tissue to which the small intestine is often compared in terms of ion and solute transporters, faces relatively small or constant osmotic gradients. Although both tissues transport water at rates over a similar range (10–70 nL cm^{-2} s^{-1}), it is interesting that while the small intestine exhibits a water permeability of 0.02 cm s^{-1}, the kidney has a permeability of 0.2 cm s^{-1} and can be accounted for by AQPs especially AQP1. The renal distribution, role, and the regulation of AQPs are well established. Thus, in addition to the constitutively expressed AQP1, AQP2 has been clearly established as the vasopressin-regulated water channel in the cortical collecting duct. Vasopressin causes both a short-term increase in the number of AQP2s in the membrane and a long-term increase in its synthesis. Although the AQPs are not critical for renal function, their presence certainly aids in salt and water transport.

6.3 UNIPORTERS AND COTRANSPORTERS

It has become apparent that AQPs and diffusion through the lipid bilayer in of themselves cannot account for the uphill transport of water observed in the intestine and other tissues and other proteins must be involved. Two important groups of membrane transporters, distinct from the AQP water channels, the uniporters and the cotransporters, have emerged as relevant to water transport. Their relevance is based on the recent demonstrations that these proteins, well-characterized as solute and ion transporters, also transport water; a cotransporter mediated "active" water transport (67, 133) was first proposed by Loo and Wright's group. In terms of process, some of them cotransport water coupled to the transport of ions, as a function of the turnover of the protein; examples include KCC on the basolateral membrane of absorptive epithelia and the NKCC1 on the basolateral membrane of secretory epithelia. Other membrane proteins utilize both protein turnover transport mechanisms and osmotic driving forces to move water across the membrane; examples include the hexose transporters, SGLT1, GLUT1, GLUT2, and GAT1. Finally, transporters like the urea channel, UT-B, use only the osmotic gradient, much like AQPs. In the latter two groups, the osmotic water permeability per protein (i.e., the amount of water transported per protein per second given a unit osmotic gradient) varies from 0.03 (10^{-14} cm^3 s^{-1}) for AQP0 to 1.4 for rabbit SGLT1, 4 for AQP1, and 7.0 for UTB.

The emerging picture on the cotransport of water has been based on observations using a variety of different approaches ranging from two-electrode voltage clamp to high-resolution volume measurements in native tissues and cell models (133). For those proteins, such as SGLT1, which also exhibit osmotic water permeability, this function appears independent of the cotransport properties and is more akin to that of AQPs. These include relatively low activation energies, ability to be blocked by specific inhibitors, and passive water permeability. In terms of the cotransport function, there are a number of key features worthy of consideration. First, the water transport capacity per protein is considerable and ranges from 590 molecules per turnover observed for NKCC1 to 40 for GLUT1. While the list of the water-conveying membrane transporters is increasing, it is noteworthy that not all isoforms of a protein transport water with the same efficacy. For example, while NKCC1 and NKCC2 transport Na^+, K^+, and 2 Cl^-, only NKCC1 transports water with each turnover of the protein. Second, the stoichiometry of the water/substrate is an inherent property of the protein and the flux of the substrates provides the driving force. The coupling ratio of the fluxes between the substrates and water are constant, regardless of the gradient (electrical, chemical, or osmotic) or rate driving the transport. For proteins that transport multiple substrates, like SGLT1, the coupling ratio decreases as the size of the substrate increases. Third, since the water transport is energized by the solute/ion gradients, it can be uphill and against an osmotic gradient and essentially is a secondary active transport mechanism. Experimental evidence has shown that even in the presence of increasing osmolarity (up to 300 mOsm), transporters such as KCC and NKCC1 can still move water. This is especially relevant in tissues facing fluctuations in luminal osmolarity, such as the intestine and gall bladder. Fourth, since the stoichiometry of water/substrate is large and the number of transporters in the membrane are governed by the requirements of the solutes/ions to be transported (e.g., stimulation of activity by regulators either by phosphorylation and/or increased incorporation of NKCC1 into the basolateral membrane in response to a secretory stimulus), these proteins could be a major contributor to the water-transporting capacity of the membrane. Finally, all of the above features together make this mode of water transport especially significant in tissues that have few or no water channels.

In contrast to the filtered pore mechanism of AQPs, the molecular basis of water transport across cotransporters or uniporters is linked to their role as substrate transporters. The binding sites for substrates in membrane proteins are situated in aqueous cavities, and both water and the substrate enter the binding site from the external surface. Upon substrate binding, there is a conformational change resulting in the binding site being exposed to the internal surface and both water and the substrate enter the inner compartment with the protein returning to its original conformation. Since the size of the aqueous cavity is limited, larger substrates would be accompanied by fewer molecules of water. It is still unclear whether these conformational changes are accompanied by local, intramembrane osmotic and hydrostatic gradients. The number of molecules of water transported as a consequence can be as high as 1320 molecules for a voltage-gated anion channel and therefore a turnover number 40–590 molecules of water reported for cotransporters are well within the range.

6.4 MECHANISMS OF WATER TRANSPORT IN THE INTESTINE

With the above background, clearly, water transport involves more than a model of passive water permeability and small osmotic gradients. There is no doubt that passive transport through AQPs, when present, is advantageous in that it increases the rate of water transport and save metabolic cost. Transepithelial water movement is function of the transepithelial osmotic gradient—thus, under isotonic conditions, when luminal and serosal osmolarity are 300 mOsm, flux of water is maximal. As the luminal osmolarity increases, flux of water decreases but does not cease until luminal osmolarity is 250–300 mOsm above that of the serosal surface. Thus, both isotonic water transport and the uphill transport of water, especially in response to a meal, have to be accounted for (Figure 12). Water can either move via a paracellular or a cellular pathway. Movement through the paracellular pathway follows passive hydrostatic and osmotic gradients and can certainly be subject to nuances of the geometry of the lateral spaces. In the cellular pathway, water has to traverse the apical membrane, the cytosol, and the basolateral membranes.

Thus, during absorption, it is most likely that in addition to passive water permeability, water enters the villus enterocyte through a variety of nutrient transporters including SGLT1 and the amino acid transporters. In addition, the recent evidence that during a meal GLUT2 also gets recruited to the membrane suggests that water can also enter through this uniporter, albeit with fewer per turnover than SGLT1 (40–110 vs. 235 for SGLT1). On the basolateral membrane, the K^+–Cl^- cotransporter is the K^+ leak pathway for the Na^+/K^+ pump with a K^+ efflux close to the K^+ entering via the pump. It couples the transport of 500 molecules of water to the energy derived from the efflux of 1 K^+:1 Cl^- and its blocker furosemide inhibits water absorption. GLUT2 is constitutively expressed in the basolateral membrane and can also contribute to water exit.

In secretory cells of the intestinal epithelium, the basolateral NKCC1 also utilizes the energy derived from the Na^+/K^+ pump to drive 1 Na^+:1 K^+:2 Cl^- into the cell along with up to 590 molecules of water. While Cl^- exits the cell through apical channels, Na follows through the paracellular pathway, and the movement of 370 molecules of water has been associated with the movement of each Na^+ and Cl^-. However, it is unclear how water exits the apical membrane in the intestine; salivary glands have AQP5 channels, and whether another AQP, such as AQP3, plays a similar role in the intestine remains to be determined.

Overall, the intestinal epithelium is complex, and multiple mechanisms are needed to ensure water homeostasis in this organ with a vastly fluctuating external milieu.

• • • •

CHAPTER 7
Regulation

7.1 INTRACELLULAR MEDIATORS

Extracellular stimuli, including hormones, neurotransmitters, and immunomodulators, need to transduce their signal within the cell for regulation of function, including transport processes, to occur. The volume of information is too vast for us to provide a detailed description of the signaling cascades, and in this chapter, we provide an overview of intracellular mediators and their extracellular signals, especially as they pertain to the intestinal epithelium.

Since the discovery of cAMP in the 1960s as an intracellular signal transducer, there has been an information explosion in the field of second messengers. Perhaps, the most important concept to convey is that there are multiple second messengers and that contrary to earlier predictions, they do not act via monolithic cascades but rather are subject to extensive cross-talk between the cascades adding to the complexity of the regulation. Most extracellular signals activate one or more of the following second messenger cascades: the cyclic nucleotides cAMP and cGMP, intracellular Ca^{2+} ($[Ca^{2+}]_i$), the inositol phosphate-diacyl glycerol, and tyrosine kinase pathways [90]. Activation of these pathways could cause short-term and rapid regulation and/or elicit longer lasting effects by altering gene transcription. Some extracellular signals, such as steroid hormones, directly modify gene transcription via specific receptors. By and large, the messenger systems are widely distributed through various phyla, and some are even present in the plant kingdom.

In mammals, structural and functional variations and tissue-, cell-, subcellular domain-, and age-specific differences abound—all of them contributing to the net biological response (Figures 12 and 13). Where does this variation come from? Signaling systems have multiple enzymatic steps, where the product of one step triggers the activity of the next protein/enzyme and so on down the cascade. In addition, many of these proteins have multiple isoforms and variants that can influence function. Most biological systems and rapid processes such as ion transport require regulation that can fluctuate minute by minute, if necessary. This is achieved by either inherent mechanisms of desensitization or through other signaling mechanisms that cause the degradation of the products of the first cascade. Thus, adenylate and guanylate cyclases catalyze the conversion of ATP and GTP to cAMP and cGMP, respectively, whereas phosphodiesterases, with different substrate specificities and themselves subject to activation, are poised to degrade the cyclic nucleotides and shut down that signaling pathway. Clearly, the biological response will be determined by the net influences of these stimulatory and attenuating pathways.

Most, if not all extracellular regulators bind to a receptor, either on the cell membrane or inside the cell, to trigger a signaling cascade. Thus, growth factors, cytokines, and inflammatory mediators activate receptor-associated tyrosine kinases, dual-specificity kinases, receptor kinases, and/or extracellular-regulated kinases (ERK) to regulate epithelial cell function. However, a majority of extracellular signals (stimulators or inhibitors) use intracellular cAMP, cGMP, or Ca^{2+} to modulate epithelial ion transport processes. Receptor isoforms and differences in receptor regulation provide the first level of variability in terms of signaling. Thus, a single regulator can turn on different intracellular cascades, depending on the receptor isoform. For example, epinephrine attenuate cAMP production when it binds to its α α receptors, but stimulates cAMP production via its β receptors.

7.1.1 Cyclic AMP

Briefly, the cAMP cascade involves at least five steps in the transduction of an external signal to a change in cell function (Figure 13). Step 1: A neurotransmitter or hormone binds to a heptahelical membrane spanning (HHMS) receptor, causing a conformational change. Step 2: This results in the activation a specific heterotrimeric GTP-binding protein by exchanging GDP for GTP binding. Step 3: The activated G proteins can either stimulate (Gαs) or inhibit (Gαi membrane-bound adenylate cyclase and alter cAMP generation from ATP. For example, vasoactive intestinal peptide (VIP) binds to one of a family of heptahelical membrane spanning receptors (HHMS, VPAC1 and VPAC2) (Figure 13) [79] to activate (Gαs) and increase cAMP, whereas somatostatin activates (Gαi and attenuates cAMP production. A step where the signal can be turned off is at the level of G protein activation; G proteins contain an inherent GTPase that returns the activated G protein returns to its nascent stage. Adenylate cyclases are multipass transmembrane proteins localized to the basolateral membrane of epithelial cells; they have multiple isoforms, which can add another level of variability to signaling. Step 4: Increases in $[cAMP]_i$ lead to cAMP binding to the regulatory subunit of the tetrameric PKA to release and thereby activate the catalytic kinase subunit. There are different PKA isoforms, and their activity can be further modulated by their binding to kinase-anchoring proteins (AKAP). Step 5: The kinase catalyzes the phosphorylation of the consensus sequences of target proteins, which can either be the transporter, such as the chloride channel or the sodium/hydrogen exchanger, or another protein that modulates transporter activity as may be the case with NKCC1 regulation. Parallel, but not identical, steps are involved in cGMP and Ca^{2+} cascades and are summarized below.

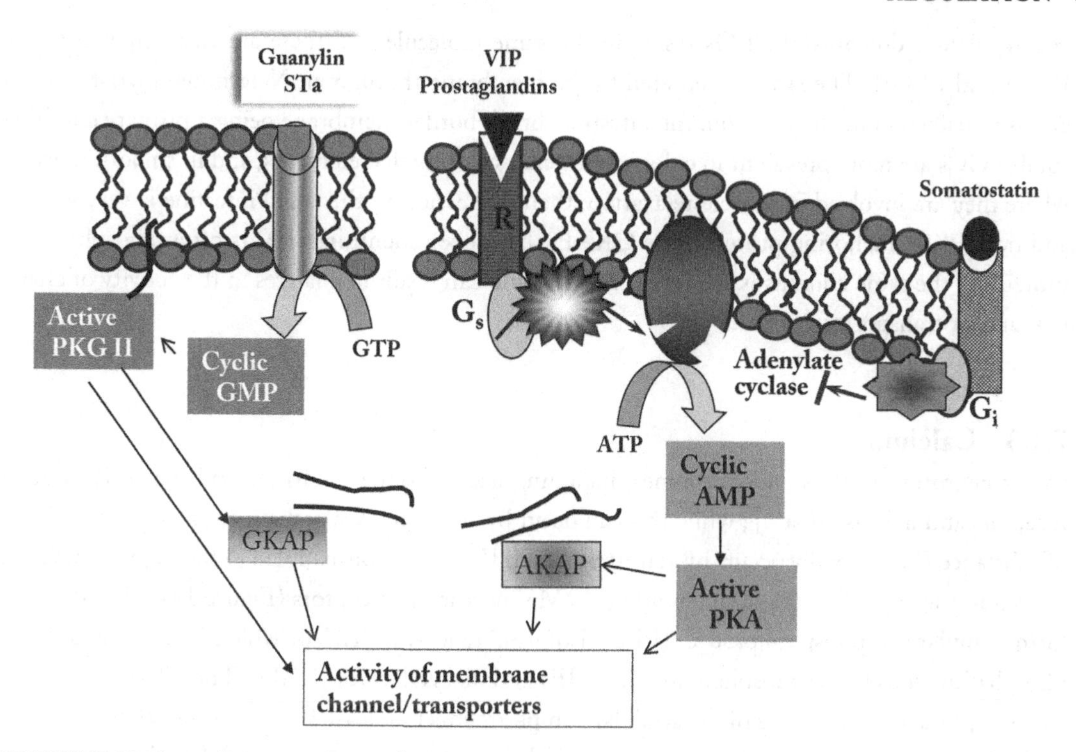

FIGURE 13: Second messengers: cyclic adenosine monophosphate (cAMP) and cyclic guanosine monophosphate (cGMP). The nucleotides are generated by specific cyclases in response to receptor activation by extracellular signals. Receptors linked to adenylate cyclase tend to be heptahelical membrane spanning (HHMS) proteins coupled to G proteins. For cGMP, the receptor binding domain and the cyclase domain reside in the same molecule. Both nucleotides activate specific kinase cascades (see Sections 7.1.1 and 7.1.2 for details). Modified from Ref. 114 (Sleisenger and Fordtran's Gastrointestinal and Liver Disease, Chapter 99; permission obtained).

7.1.2 Cyclic GMP

In contrast to cAMP, cGMP can be generated either by the activation of membrane guanylate cyclases (GCs), as seen for the atrial natriuretic peptides, guanylin, uroguanylin, and the heat-stable enterotoxins or by activation of soluble GCs as seen in the case of nitric oxide (Figure 13). Distinct from the adenylate cyclases, membrane GCs have a single transmembrane domain and serve the dual function of receptors on their extracellular domain and as an enzyme on their intracellular domain catalyzing the conversion of GTP to cGMP. Thus, unlike the adenylate cyclases, which can couple to different receptors, the ligands (atrial natriuretic peptide, guanylin, and uroguanylin) for the GC isoforms are specific. The increases in $[cGMP]_i$ by either membrane or soluble GCs results in the activation of dimeric cGMP protein kinases (PKG). Although related to PKAs, the catalytic and regulatory domains of PKGs reside in the same molecule and there are two major isoforms PKG1 and PKGII. The latter is tethered to the membrane through an N-terminal myristoylation and is restricted in its distribution, the intestinal brush border membrane being a primary site. The soluble GCs are more prevalent in colonic epithelia, subepithelial elements, and in smooth muscles where they are involved in muscle relaxation; their expression in the small intestine is very low. In addition, PKG-anchoring proteins, the GKAPs, have been identified. PKGs often phosphorylate proteins at the same consensus sequences as PKA and can result in changes in the activity of channels and/or transporters.

7.1.3 Calcium

Other neurotransmitters and hormones, including acetylcholine (Ach) and substance P, bind to receptors and increase $[Ca^{2+}]_i$; while this can occur by activating Ca^{2+} channels, as seen in the case of substance P, it generally occurs by activation of HHMS receptors coupled to the G protein, $G\alpha q$, as seen in the case of acetylcholine binding to M3 muscarinic receptors (Figure 14). The activated $G\alpha q$ stimulates a phospholipase C (PLC) isoform, generally PLC-β, which in turn hydrolyzes phosphatidyl inositol bisphosphate to release IP_3 and diacylglycerol (DAG). The IP_3 binds to specific receptors on membranes of intracellular compartments to release Ca^{2+}. This increase in $[Ca^{2+}]_i$ can activate a variety of biological processes including ion transport (see below). On the other hand, DAG does not increase $[Ca^{2+}]_i$ but causes a rapid activation of one or isoforms of PKC, a family of phospholipid-dependent, serine–threonine kinases that have multiple biological actions, including affecting ion transport. It is also important to note that DAG can also be generated independent of IP_3 from phosphatidic acid by the activation of phospholipase D by tyrosine kinase receptors.

It is essential for the cell to tightly regulate free $[Ca^{2+}]_i$ and maintain it at submicromolar concentrations despite the 1–2 mM concentration in the plasma. Prolonged elevation of free $[Ca^{2+}]_i$ is detrimental to cell function and can result in the activation of proteases, mitochondrial enzymes, and cell death. However, a transient elevation in $[Ca^{2+}]_i$ is an immensely useful signal and can either directly activate target proteins, such as Ca^{2+} channels, or elicit its effects by binding to the ubiquitous Ca^{2+}-binding protein calmodulin, and the Ca^{2+}–calmodulin complex can activate other proteins including specific calcium–calmodulin protein kinases. The cell has a number of mechanisms to ensure that calcium levels return to baseline quickly. These include desensitization to receptor stimuli, efflux via Na^+/Ca^{2+} exchange on the plasma membrane, or a rapid resequestration by Ca^{2+}-dependent ATPases on the endoplasmic reticulum. The transient receptor potential channels (TRPCs) allow for the replenishment of $[Ca^{2+}]_i$ from the extracellular compartment. The transient nature of this signaling system make Ca^{2+}-dependent secretagogues the prime candidates for the minute-by-minute regulation needed in the intestine. Another level of control is exercised by the PLCs; along with IP_3 and DAG, they may release polyinositol phosphates, such as inositol

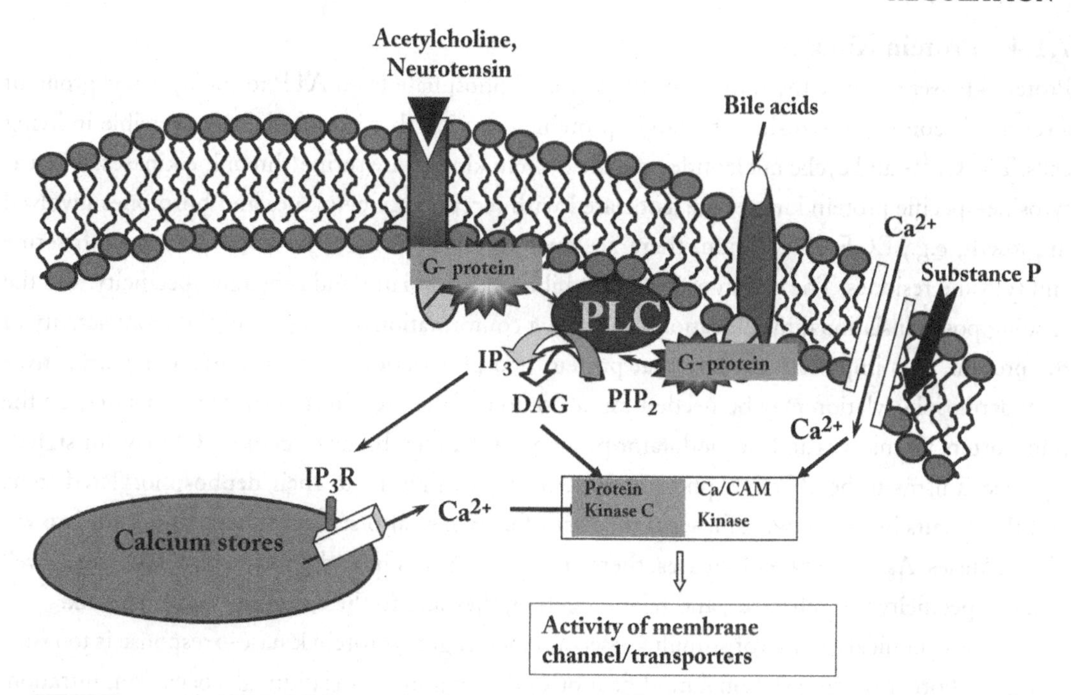

FIGURE 14: Second messengers: calcium. Hormones and neurotransmitters such as substance P and acetylcholine, activate secretion by increasing $[Ca^{2+}]_i$. Substance P can stimulate Ca^{2+} channel activity, and acetylcholine binds to M3 muscarinic HHMS receptors coupled to the $G\alpha q$ class of G proteins to activate phospholipase C and generate IP_3 (see Section 7.1.3 for details of this signaling pathway). Modified from Ref. 114 (Sleisenger and Fordtran's Gastrointestinal and Liver Disease, Chapter 99; permission obtained).

3, 4, 5, 6 tetrakisphosphate, which function as ileal brakes and dampens Ca^{2+}-induced Cl^- secretion. Phenothiazines and loperamide may interfere with Ca^{2+} signaling.

An increase in any one of these second messengers results in an increase in fluid secretion, and the activation of any one or more of the underlying transporters (Tables 2 and 3). Thus, cAMP could increase the activity of basolateral K^+ conductance and NKCC1 and of apical Cl^- and K^+ channels, thereby increasing electrogenic Cl^- secretion. It can also inhibit the apical NHE, NHE3 associated with Na^+ absorption. In addition to affecting the activity of the transporter, cAMP is also known to promote trafficking of transporter-bearing vesicles (CFTR in crypts and distal colonic Na^+ channels) to the apical membrane, thereby increasing the number of specific transporters in the membrane [7, 91]. The net result is an increase in luminal fluid in response to cAMP. It therefore follows that agents that promote fluid absorption are associated with a decrease in cAMP, cGMP, and Ca^{2+}; in addition, activation of some tyrosine kinase pathways leads to absorption.

7.1.4 Protein Kinases

Protein kinases catalyze the transfer of the terminal phosphate from ATP to the hydroxyl group of a serine, threonine, or tyrosine of a target protein, an action that is essentially irreversible in living cells. The Ca^{2+}- and cyclic nucleotide-specific protein kinases are serine/threonine kinases, whereas tyrosine-specific protein kinases are associated with receptors of cytokines and hormones involved in growth, e.g., EGF. In addition, there are dual-specificity kinases phosphorylating both serine and tyrosine residues. The action of kinases exhibit both regulator and substrate specificity, and the ensuing posttranslational modification results in a conformational change and therefore activity of the protein. It is important to note that protein phosphorylation is not synonymous with activation; dephosphorylation may be needed for activation of the protein. Phosphoproteins may be the transporters themselves and/or modulator proteins in the membrane or cytosol. Clearly, for signaling mechanisms to be effective, phosphoproteins need to return to their dephosphorylated state, and this occurs by the action of a separate class of enzymes, also subject to regulation, the protein phosphatases. As in the case of kinases, there are specific Ser–Thr and tyrosine phosphatases as well as dual-specificity phosphatases, and taken together, they add to the complexity of cross-talk.

The canonical cascade of stimulus→second messenger→protein kinase→response is too simplistic, and other posttranslational modifications including myristoylation, glycosylation, nitration, and sumolyation need to be taken into consideration. In addition, cross-talk with other signaling cascades and, much like classic enzymology, feed forward and feedback regulatory steps need to be considered.

7.1.5 Compartmentalization

Another cellular feature especially relevant to polarized epithelial cells is the intricate compartmentalization of the transporters and components of the signaling cascade; such localization to subcellular or membrane domains is the new norm. This occurs either via scaffolding proteins interacting with cytoskeletal microfilaments, intermediate filaments, or microtubules or via sequestration in vesicles. These processes are not static but are highly dynamic and subject to regulation. At the level of the membrane, there are cholesterol-rich membrane domains such as lipid rafts, which influence membrane fluidity and therefore anchor specific transporters and their regulators. Many anchoring domains serve as multienzyme signaling complexes. Thus, cytosolic scaffolding proteins serve as docking sites, through PDZ domain or AKAP type interaction, for various components of the signaling system including cyclases, kinases, phosphatases, and phosphodiesterases. For example, GCC and the intestinal PK GII have cytoskeletal and membrane interacting domains that bring them in close to CFTR in the brush border membrane (Figures 10 and 13). A very effective way to increase activity is to alter the V_{max} of the transporter and is generally achieved by increasing active transporters in the membrane. Rather than rev up the cell's machinery to generate new proteins, this is achieved by trafficking of transporters into and out of the membrane via endosomal vesicles. For example, in the intestinal epithelium, increases in cAMP simultaneously promote NHE3 retrieval from the membrane by endocytosis and CFTR translocation to the membrane by exocytosis, thereby causing an increase in luminal fluid accumulation by inhibiting Na^+ absorption and stimulating Cl^- secretion (Figure 7). The study of CF has underscored the centrality of the cell's quality control systems; thus, the most common mutation in CF, the deletion of a single amino acid, phenylalanine 508, results in this 1400+ amino acid protein being misfolded, ubiquitinylated, and tagged for degradation in the proteasomes (Figure 10).

7.1.6 Transcriptional and Translational Control

Not all transporter activation is rapid, and long-term alterations by regulation at the transcriptional level are also needed and known to occur. The best example is the increase in colonic ENaC synthesis in response to aldosterone. In addition, ENaC is subject to rapid activation by signaling systems. A method of regulation that has emerged in the past 5–10 years is the role of microRNAs (miRNA). miRNAs are short RNA transcripts and are an endogenous mechanism of regulating gene translation. They are transcribed from DNA like other RNAs, and then cleaved by the protein complex Drosha within the nucleus. Following export into the cytoplasm, the miRNA is further cleaved by Dicer, forming mature miRNA about 22 nucleotides in length. miRNAs regulate post-transcriptionally by binding to complementary sequences of target mRNA transcripts and can have several target mRNAs. Depending on partial or exact complementarity to the target mRNA, base-pairing can result in inhibition of translation or mRNA degradation. In humans, miRNAs bind to the 3' untranslated region (UTR) of mRNA. These micromanagers of protein output are well conserved and thought to play a role in evolution of gene regulation. In the intestine, the inflammatory cytokine TNFα results in an increase in the expression of miRNA 122a. This miRNA binds to the UTR of occludin mRNA, leading to its degradation. Depletion of occludin in enterocytes results in an increase of intestinal TJ permeability [131].

Finally, it must be noted that to fully grasp the physiological relevance of the complexities of the signal transduction networks, they need to be assessed in the intact intestine, a daunting challenge indeed.

7.2 PINES: PARACRINE-IMMUNO-NEURO-ENDOCRINE SYSTEM

The major neurohumoral substances and toxins that regulate intestinal fluid transport are provided in Tables 1, 2 and 3. Agents that regulate net fluid absorption increase Na^+ uptake and attenuate Cl^- secretion, whereas those that promote net fluid secretion generally inhibit Na^+ absorption and stimulate Cl^- secretion. Although the healthy gut has patent electrolyte absorptive and secretory

TABLE 1: Agents stimulating intestinal absorption
ENDOGENOUS AGENTS
α-Adrenergic agonists
Aldosterone
Angiotensin
Enkephalins
Glucocorticoids
Growth hormone
Neuropeptide Y
Peptide YY
Prolactin
Short-chain fatty acids
Somatostatin
PHARMACOLOGIC AGENTS
Berberine
Clonidine (α_2-agonist)
Cyclooxygenase inhibitors
Glucocorticoids
Lithium
Mineralocorticoids
Octreotide
Opiates
Propranolol

TABLE 2: Endogenous agents stimulating intestinal secretion

SOURCE	AONGIST	INTRACELLULAR MEDIATOR
Immune	Adenosine	cAMP
Immune, cell membranes	Arachidonic acid	cAMP
Immune	Bradykinin	cAMP
ENS	Peptide histadine isoleucine	cAMP
Immune	Platelet activating factor	cAMP
Immune, mesenchymal	Prostaglandins	cAMP
Immune	Reactive oxygen metabolites	cAMP
Paracrine	Secretin	cAMP
ENS/VIPoma	Vasoactive intestinal polypeptide	cAMP
ENS	Acetylcholine	Ca^{2+}
??	Bombesin	Ca^{2+}
	Galanin	Ca^{2+}
Immune	Histamine	Ca^{2+}
Endocrine M cells	Motilin	Ca^{2+}
ENS	Neurotensin	Ca^{2+}
ENS, carcinoid	Serotonin	Ca^{2+}
ENS, carcinoid	Substance P	Ca^{2+}
Heart	Atrial natriuretic peptide	cGMP
Goblet, epithelial cells	Guanylin	cGMP
Immune, mesenchymal	Nitric oxide	cGMP

TABLE 2: Endogenous agents stimulating intestinal secretion (*continued*)

SOURCE	AONGIST	INTRACELLULAR MEDIATOR
ENS, MCT	Calcitonin, calcitonin gene-related peptide	?
??	Gastric inhibitory polypeptide	?
Paracrine	Gastrin	Ca^{2+} (PKC/MaPK?)
Immune	Leukotrienes	?

TABLE 3: Luminal agents stimulating intestinal secretion

AGENT	ACCESSORY PATHWAY	INTRACELLULAR MEDIATOR
Bacterial Enterotoxins		
Accessory cholera enterotoxin		??
Aeromonas enterotoxin		cAMP
Campylobacter jejuni enterotoxin	?	cAMP
Clostridium difficile (toxin A)	Cytoskeleton	Ca^{2+}
Clostridium perfringens		??
Escherichia coli (heat stable toxin)	ENS	cGMP
Escherichia coli (heat labile toxin)	ENS	cAMP
NSP4	Cytoskeleton	Ca^{2+}

TABLE 3: Luminal agents stimulating intestinal secretion (*continued*)

AGENT	ACCESSORY PATHWAY	INTRACELLULAR MEDIATOR
Rotavirus		
Salmonella enterotoxin	?	cAMP
Vibrio cholerae enterotoxin	ENS, 5-HT	cAMP
Vibrio parahaemolyticus		Ca^{2+}
Yersinia enterocolitica	?	cGMP
Zona occludens toxin (ZOT)	Cytoskeleton	??
Miscellaneous		
Bile salts		cAMP/Ca^{2+}
Laxatives		??
Long-chain fatty acids		cAMP/Ca^{2+}

processes, the net result is absorption, accompanied by the expulsion of waste material as "healthy" stools. Disruption of this balance can cause diarrhea or, in some cases, constipation, and therefore, the balance needs to be continuously and well-regulated.

Factors that regulate intestinal ion transport emanate from a variety of sources, starting with the luminal intake, enterocyte per se (autocrine mediation), and from classic paracrine, immunologic, neural, and endocrine systems (PINES) (Figure 15). The classical delineations of these categories has blurred over the past half century, as the evidence that regulators of one category can also be found in others. For example, CCK, long thought to be a classical endocrine hormone, is present in nerve endings and has a significant effect as a neural mediator. Aberrations in any one of the PINES pathways or in their interactions can contribute to disease states. In the lumen, the physical presence of food and motility patterns can lead to epithelial stroking and stretching of the intestinal wall. This could activate mechanoreceptors in the nerve endings, which in turn lead to the stimulation of one or more branches of the PINES. On the other hand, luminal chemical stimuli activate chemoreceptors, which, in turn, can trigger other or overlapping aspects of PINES. The

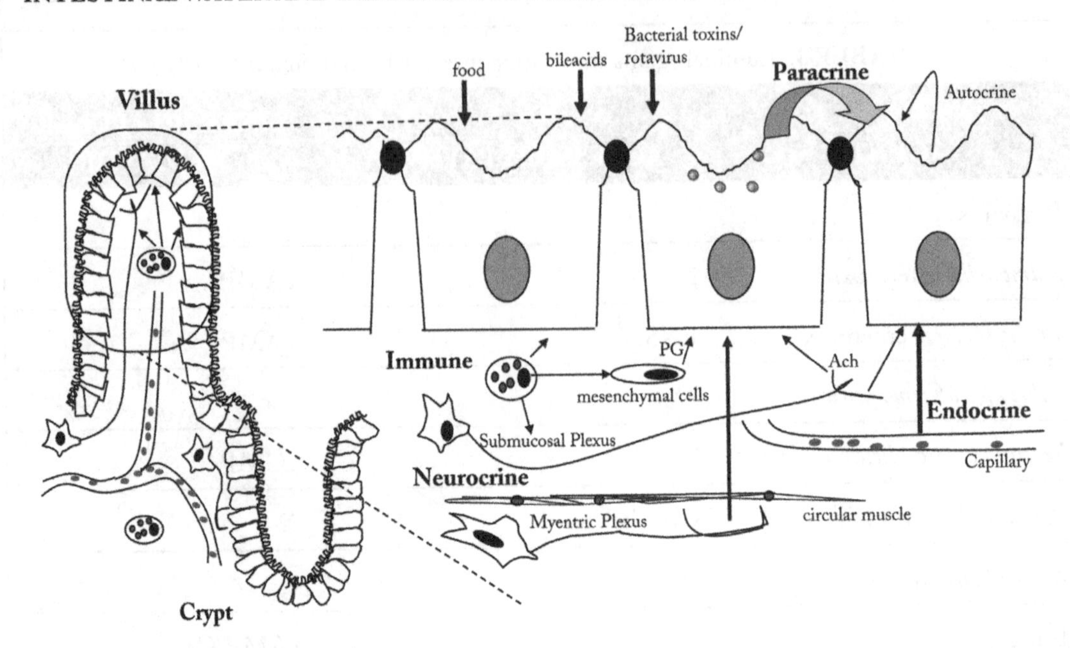

FIGURE 15: Paracrine, immunologic, neural, endocrine system (PINES): intestinal ion transport is regulated by all aspects of PINES and there is extensive cross-talk among these regulatory systems. Individual modulators can either act directly on the epithelial cell or indirectly influence other regulators to modify transport. For example, components of the neural network interact with blood vessels, immune cells, enterochromaffin cells, paracrine cells, enterocytes, and other neurons. They could respond to luminal contents and stretch, by releasing acetylcholine to stimulate Cl^- secretion or to alter muscle motility (see Section 7.1 for details). Modified from Ref. 114 (Sleisenger and Fordtran's Gastrointestinal and Liver Disease, Chapter 99; permission obtained).

structural elements of PINES are close to the subepithelial space. Thus, activation of neurons could easily trigger release of mediators from mast cells that are in the vicinity. As noted in the previous section, the net effect of these interactions will be compounded by the multiple array of receptors and signaling pathways. It is easy to see how such an intricate network of signaling has many aspects and can be responsible for the minute-by-minute regulation needed for the intestine to function properly (Figure 15) and as is evidenced by aberrations seen in disease states. A few examples are provided here: cholera toxin directly acts on epithelial cells while simultaneously stimulating neural, paracrine, and immune responses distal from the site of exposure (Figure 16). In another example, a single neurocrine substance can act via specific receptors to affect function in a number of cell types as exemplified by serotonin (5-hydroxytryptamine) released by mucosal enterochromaffin cells. It acts on the myenteric neurons to release Ach and stimulate migratory contractions; it acts on sub-

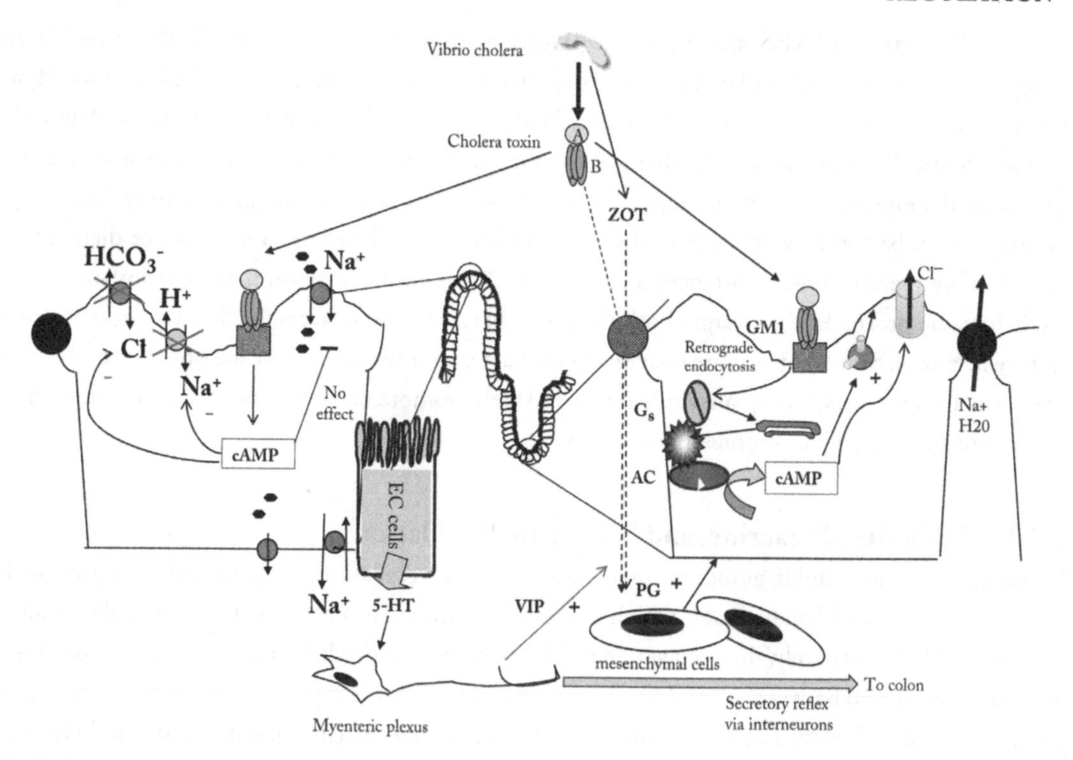

FIGURE 16: Cellular actions of *Vibrio cholerae. V. cholerae* produces an enterotoxin, CT, and a zonula occludens toxin (ZOT) that disrupts tight junction permeability. CT binds via its B subunit to GM1 gangliosides (GMI) on the apical membrane and inserts the A subunit into the cell. The A1 subunit ADP ribosylates Gαs and irreversibly activates adenylate cyclase to produce cAMP. Cyclic AMP stimulates Cl^- secretion in the crypt epithelial cells (cell on the right side) and inhibits Na^+ absorption via NHE, but not via the Na^+-glucose cotransporter in the apical membrane of the villus surface epithelial cells (cell on the left side). Access of CT to the subepithelial space via disrupted permeability results in the activation of PINES elements, including secretory reflexes to the colon via interneurons, to exacerbate secretion (see Chapters 7 and 8 for details). Modified from Ref. 114 (Sleisenger and Fordtran's Gastrointestinal and Liver Disease, Chapter 99; permission obtained).

mucosal neurons to release Ach and calcitonin gene-related peptide to stimulate peristalsis and secretory reflexes and directly on epithelial cells to stimulate secretion. A third example is that of VIP; in the healthy adult, VIP is a peptidergic neurotransmitter that stimulates exocrine gland secretion, including in the intestine, and causes smooth muscle relaxation. Yet, Verner-Morrison syndrome, also known as pancreatic cholera, is classified as an "endocrine-mediated" diarrhea, since pancreatic islet cell tumors produce copious amounts of VIP.

All aspects of PINES are important players in the defensive response of the intestine to injury, including bacterial onslaught. A luminal challenge to the intestine could alter blood flow, motility, mucous production, and, of course, fluid secretion, triggering multiple interrelated responses. Some diarrheas are due to changes in motility. Thus, small-volume diarrhea is associated with altered anorectal motility, and rapid intestinal transit is observed post-gastrectomy. Decreased motility causes bacterial overgrowth in the small intestine, and the bacteria may induce diarrhea by one of many mechanisms. Substances associated with the immune system, such as inflammatory mediators and prostaglandins, target both the epithelial and muscle layers to elicit a coordinated secretory response. Similar effects are observed with bacterial enterotoxins. At the opposite end of the spectrum are agents such as opiates/enkephalins, which promote net absorption by both stimulating electrolyte absorption and suppressing motility.

7.2.1 Autocrine, Paracrine, and Juxtacrine Regulation

Epithelial cells can exhibit autocrine regulation (i.e., self-regulation) by elaborating factors such as prostaglandins and leukotrienes, which can bind to surface receptors of the same cells to alter function [89]. Alternatively, they can act on neighboring cells and exhibit paracrine regulation (Figure 15). Interspersed between enterocytes in the epithelium are the intestinal endocrine cells, which are histologically identified as the enterochromaffin cells. Sensors (mechanoreceptors and chemoreceptors) on the apical membrane of these cells react to changes in the luminal environment by rapidly releasing, from their basolateral surface, secretory granules full of biogenic amines and hormones. These hormones can either act in a local (paracrine) manner by affecting neighboring cells in the gut wall or traverse the basement membrane and/or enter the systemic circulation to act on distant target cells in a classic endocrine manner (see Section 7.2.4). Finally, another term applied to extracellular messengers is *juxtacrine mediators*. Although, less frequently used, this terminology is apt for describing the regulatory products of cells that are not traditionally defined as endocrine cells such as mesenchymal cells and that affect neighboring cells. Intestinal myofibroblasts are a rich source of chemokines, cytokines, eicosanoids, and growth factors, all of which can modulate intestinal transport.

7.2.2 Immunologic Regulation

The lamina propria underlying the epithelial layer houses a specialized tissue, the GALT, which has immunocompetent cells. While sharing some common features with the systemic immune system, GALT cells have other distinguishing features. Clearly, the composition of the different immune cells is altered during inflammation.

In the healthy gut, the composition of the immune cells are as follows: T lymphocytes, 60%; B lymphocytes and plasma cells, 25% to 30%; macrophages, 8% to 10%; mast cells and polymorphonuclear cells (usually eosinophils), 2% to 5% [101]. They secrete chemokines, cytokines, eicosanoids, nucleotides, biogenic amines, many of which they share with other branches of PINES. During inflammation, the number of immunocytes, is increased and the pattern of distribution is related to the cause of inflammation (see Chapter 8), and it therefore follows that the type of cytokines elaborated and their effects on motility and transport will vary.

Products of immunocompetent cells ranging from oxidants and eicosanoids to peptides influence ion transport by altering barrier function, acting on the enterocyte directly to alter ion transport processes, or indirectly influencing intestinal function by modulating other parts of the PINES network (Figure 15). Examples include oxidants such as superoxides; eicosanoids such as arachidonic acid and its metabolites, leukotrienes and prostaglandins; and peptides such as the kallikreins and bradykinin, cytokines, interferon-γ, platelet-activating factor, and substance P (which is also a peptidergic neurotransmitter) (Figure 15). Key regulators in inflammation are the prostaglandins; mediators such as bradykinin liberate arachidonic acid to stimulate prostaglandin production; prostaglandins, in turn, can act directly on epithelial cells as well as interact with enteric neurons through one of many specific receptors leading to multiple responses. It is important to recognize that while we can characterize the complex interactions of inflammatory mediators in either healthy tissue or in reductionist systems, cells damaged by the inflammatory process in vivo may not show a similar profile of responses. (see Chapter 8).

7.2.3 Neural Regulation

The labyrinthine enteric nervous system (ENS) is the end controller of neural input into the intestine, and in addition to its own local circuitry, it can modulate the participation of the parasympathetic and sympathetic divisions of the autonomic nervous systems as well (Figure 14). The ENS influences all aspects of intestinal function ranging from motility in the musculature to vascular properties and last but not least epithelial cell function. It has long been recognized that the two branches of the autonomic system differentially regulate intestinal ion transport. Adrenergic stimulation, through the prevertebral and sympathetic ganglia, promotes absorption in most segments of the intestine, whereas cholinergic stimulation through parasympathetic vagal input predominantly stimulates secretion. While it has been suggested that there may be specific mechanosensory neurons resident in the ENS, the prevailing theory is that the neurons that function truly as sensory neurons have their cell bodies in the dorsal root ganglia and nodose/jugular ganglia [125].

The number and variety of neuroactive agents in the ENS parallel those in the brain and function with equal complexity involving multiple pathways. Thus, rather than release a single class of neurotransmitters, individual neurons may release specific combinations of mediators, e.g., ATP, CCK, gastrin-releasing peptide, VIP. These mediators may either act as classic neurotransmitters or as neuromodulators, fine-tuning the neuronal circuits of presynaptic sites of neurons. They can also have biphasic effects depending on the concentrations released or, as in the case of serotonin, can function as paracrine substances.

Like other neural networks, the ENS has reflexes, with important clinical implications. Target cells for neurons include components of PINES, blood vessels, and, of course, epithelial cells. Sensory input into the ENS comes from changes in the luminal content (e.g., acidity, dietary content, pathogens) or volume (e.g., stretch). Thus, acid or distension can activate TRVP1 vanilloid receptor on capsaicin-sensitive afferent nerves; in turn, these causes vasodilation by directly activating submucosal arterioles and evoke secretion by stimulating submucosal neurons [42, 113]. Dendritic, endocrine, dendritic, and/or paracrine cells releasing serotonin, adenosine, and other signals are implicated as auxiliary sensors. Cholinergic interneurons are believed to underlie the ENS mediated regulation of colonic epithelial responses to distant small intestinal challenges. Secretomotor neurons that innervate epithelial and submucosal cells can be cholinergic or VIPergic, each releasing additional neuroactive substances. A basal cholinergic secretory drive can be tempered by the sympathetic tone; thus, the loss of adrenergic sympathetic innervation in diabetic neuropathy can lead to the development of "diabetic diarrhea," and administration of α_2-adrenergic agonists can help alleviate these symptoms [18].

7.2.4 Endocrine Regulation

With the caveat that the demarcation between endocrine and other members of the PINES network is increasingly blurred, below we provide you a description of regulators that play important roles in the intestine and are often classified as "endocrine" (i.e., secreted at a site and transmitted through the systemic circulation to the target site of action) substances.

7.2.4.1 Absorptive Factors. Physiologic absorptive mechanisms are necessary to ensure the normal functioning of the intestine and to counterbalance fluid loss. Table 1 lists intestinal agents that stimulate absorption. The group of compounds that both inhibit active secretion and enhance active absorption by directly acting on enterocytes include steroid hormones (see below), catecholamines such as norepinephrine and dopamine and peptide hormones, such as neuropeptide Y, enkephalins, and somatostatin. All these peptides are also present in nerve endings and somatostatin also serves as a paracrine modulator. The catecholamines and peptides all act via heptahelical membrane receptors to activate $G\alpha_i$, thereby suppressing the cAMP pathway, to stimulate Na^+ absorption and decrease Cl^- and HCO_3^- secretion. While epinephrine activation of α-adrenergic receptors activates $G\alpha_s$ and increase cAMP in the intestine, when epinephrine binds to α-adrenergic receptors, it activate $G\alpha_i$ to stimulate absorption. This is the basis for the development of clonidine as an antidiarrheal agent, to treat diabetic diarrhea.

A major proabsorptive hormone of the small intestine is neuropeptide Y (NPY) or peptide YY (PYY). It is a 36 amino acid peptide which is predominantly found in the crypts and neuroendocrine cells, and G protein coupled membrane receptors for PYY are expressed in human and rat gastrointestinal epithelia. In association with digestion, PYY acts as an intestinal "brake" by inhibiting motility, Cl^- secretion as well as pancreatic and gastric secretions and increasing contact time of nutrients with the intestinal mucosa to promote absorption [122].

Somatostatin, produced by the enteroendocrine D cells, stimulates fluid and electrolyte absoroption in the intestine and inhibits the secretory effects of other agents. Somatostatin acts via multiple ways to inhibit secretion; decreases paraneoplastic effects of tumors, slows transit time, and directly affects the epithelial cell. Octreotide, a long-acting analog of somatostatin, is used effectively to treat many endocrine related diarrheal diseases. Other peptide hormones, such as angiotensin II and insulin, have been implicated as pro-absorptive agents but their physiologic significance is as yet unknown.

Another important group of antidiarrheal agents that act by decreasing secretion and inhibiting motility are the opioid peptides. The use of plant opiates to treat diarrhea dates back two millennia to the early Egyptians ong before its analgesic properties were recognized. In the 1970's, receptors for opiates were discovered and the major receptor subtypes μ, δ, and κ were identified. Curiosity of why mammalian systems had receptors for plant alkaloids led to one of the early and fascinating discoveries of "molecular mimicry." A family of endogenous mammalian opioid peptides including enkephalins, endorphins and dynorphins that specifically bind to the opioid receptors were isolated and characterized. Acting via one of three main opioid receptors, these opioid peptides can either act directly on the epithelial cells, smooth muscle cells or modify the electrical and synaptic behavior of ENS neurons. Thus, the constipation associated with morphine intake could be due to its effect on the ENS and suppression of secretion or by a stimulation of sympathetic noradrenergic discharge. A G protein coupled pathway that activates K^+ channels and inhibits Ca^{2+} channels has been shown to be the basis of these effects. Constipation is one of the side effects that needs to be managed in patients receiving opiates for pain.

7.2.4.2 Steroid Hormones. In many epithelia, expression of proteins participating in vectorial ion transport such as the Na^+-K^+-ATPase and ENaC are up-regulated by glucocorticoid and mineralocorticoid hormones, thus leading to enhanced Na^+absorption and K^+ secretion. Glucocorticoid and mineralocorticoid receptors are present in proximal and distal colon in both surface and crypt epithelia. These steroids exert differential effects on Na^+ absorption in the colon, with a clear segmental heterogeneity in a variety of species, including humans (see Chapter 4). In the proximal colon, Na^+ transport is primarily mediated by an electroneutral process, and despite the presence of mineralocorticoid receptors, Na^+ absorption is not influenced by aldosterone. Similar to the proximal colon, mineralocorticoids have no effect on the small intestinal electroneutral absorption. There are distinct species variations with respect to the distal colon. Thus, in the rat, electroneutral absorption is the predominant transport process in both proximal and distal colon in the absence of aldosterone. However, in other species including human, Na^+ transport in the distal colon is dominated by electrogenic absorption via amiloride-sensitive Na^+ channels, which are under the influence of exquisite mineralocorticoid, i.e., aldosterone, regulation. Aldosterone action could also be divided into early and late responses, with early effects being attributed to a nongenomic action (increase activity of ENaC) while late responses include genomic actions. Aldosterone increases Na^+ absorption by increasing the number and activity of ENaC (see Chapter 5) and the Na^+/K^+ pump, via serum regulated kinase, SGK1. The early action of aldosterone involves SGK1, and the enzyme is up-regulated in the kidney collecting ducts and in the colon in response to aldosterone (also see below). Aldosterone not only regulates Na^+ transport, in the rat distal colon, it also converts K^+ absorption to net secretion by binding to intracellular type I mineralocoritcoid receptors and stimulates the genomic effects. It has been demonstrated that an increase in pH via an up-regulation of NHE is responsible for the activation of the K^+ channels and that it involves a rapid regulatory effect of aldosterone on electrolyte transport. Further, this fast and nongenomic action of aldosterone on Na^+/H^+ exchange in the rat distal colon occurs via an increase in intracellular $[Ca^{2+}]_i$ and stimulation of PKC. Aldosterone has also been demonstrated to increase K^+ absorption and secretion. There seems to be developmental differences in Na^+ absorption in the colon, with it being higher in neonates than in adult and this difference is due to high levels of aldosterone in the neonates. The clinical significance of aldosterone in regulation ion transport is seen in patients with Addison's disease, where aldosterone deficiency causes diarrhea.

Low-dose glucocorticoids induce electroneutral Na^+ absorption in both proximal and distal colon, whereas high concentrations activate both electroneutral and electrogenic absorption, particularly in the distal colon of human and rat. One should keep in mind that due to the high glucocorticoid concentration and promiscuity of mineralocorticoid receptors, corticosterone could mimic aldosterone regulation in the intestine. It has been shown that up-regulation of electroneutral Na^+ absorption by glucocorticoids is paralleled by inhibition of electrogenic absorption. Like mineralocorticoids, glucocorticoids control epithelial Na^+ conductance by activation of transcription of ENaC subunits,and at the level of apical transporters, they also regulate activity of the Na^+/K^+ pump. With regard to electroneutral Na^+ absorption, glucocorticoids regulation is isoform-specific. Thus, in the rabbit intestine and colon, glucocorticoid induced increases in Na^+ absorption is associated with increases in NHE3, but not NHE2 or NHE1, mRNA and protein [132]. Serum and glucocorticoid regulate kinase SGK, a member of the serine-threonine kinase family. Early responses to glucocorticoids are shown to involve the activation of SGK1 and late responses are via transcriptional modification of transporters and channel proteins. As a result, of these pro-absorptive effects, agonists of glucocorticoid receptors have been widely used in the treatment of diarrheal diseases.

7.2.4.3 Secretory Factors. In this section the roles of some hormones responsible for stimulating fluid secretion in the intestinal lumen will be discussed (see Table 2). These can be broadly classified into four categories: i) inflammatory agents such as prostaglandins, leukotrienes and histamine; ii) neurotransmitters, such as VIP, acetylcholine, nucleotides, Substance P; iii) paracrine agents such as serotonin and neurotensin; iv) and members of the guanylin family such as guanylin and uroguanylin, and peptides released by the intestinal goblet and epithelial cells.

Eicosanoids are a family of 20-carbon oxygenated metabolites of arachidonic acid which originates from membrane phospholipids and are produced primarily in the subepithelium. Metabolism of arachidonic acid by cyclooxyegenase produces prostaglandins and by lipoxygenase produces leukotrienes. All these agents inhibit electroneural NaCl absorption and stimulate electrogenic Cl^- secretion.

Prostaglandins, primarily arising from inflammatory cells in the submucosa, have autocrine actions in epithelial cells. Prostaglandin effects are multifactorial, involving modulation of the enteric nerves, direct action on epithelial cells, altering smooth muscle function and thereby affecting intestinal motility and blood flow. Prostaglandins primarily act via cyclic AMP, and to a smaller extent via $[Ca^{2+}]_i$. Cyclooxyegenase inhibitors, such as indomethacin or aspirin, increase basal rates of absorption, implying that prostaglandins contribute to the basal secretory tone of the epithelium. In inflammatory bowel disease, there is increased intestinal production of eicosanoids which contributes to diarrhea. Glucocorticoids decrease prostaglandin secretion and hence are used to treat inflammation. Futher, the 5-amino-salicylates (5-ASA or mesalamine) and their sulfa derivative, (sulfasalzine), are more effective than traditional inhibitors of cyclooxygenases, such as acetyl salicylic acid (aspirin) in treatment of inflammatory bowel disease. Leukotrienes are considered to also have a role although the mechanism of their action is less well understood but may involve activation of secretomotor neurons in the subepithelium.

Acetycholine, serotonin, VIP, guanylin, other hormones and neurotransmitters stimulate intestinal secretion by inhibiting electroneutral NaCl absorption and stimulating Cl^- secretion, and they all play an important role in the complex physiology of the gut. They are classified below by their mechanisms of action. Acetylcholine, acts via the muscarinic M3 receptor, serotonin and neurotensin increase $[Ca^{2+}]_i$, whereas VIP, and related peptide hormones such as secretin, increase intracellular cyclic AMP, and guanylin increases cell cGMP (see below).

VIP is a 20-amino acid peptide, structurally related to another gastrointestinal hormone, secretin. VIP has substantial homology with pituitary adenylate cyclase-activating peptide (PACAP), a 27-amino acid protein which predominates in the intestine. The prosecretory effect of these peptides are due to their effect on the smooth muscles and hence increasing motility. Further, VIP mediates some of the secretory effects of 5-HT as a VIP receptor antagonist inhibits the secretory effect elicited in vitro by a 5-HT3 receptor agonist (see section on serotonin).

Substance P belongs to the tachykinin family and is a short peptide with 11 amino acids. It has been recognized for its secretory properties since the 1930's. It is widely distributed in the gastrointestinal tract, primarily in the small intestine and alters blood flow and intestinal motility. Like other agents, kinins act on the basolateral membrane Na^+/K^+ ATPase and the apical CFTR chloride channel. Furthermore, kinins have been shown to mediate some of the secretory effects of cholera toxin-induced secretion in the rat intestine.

Serotonin (5-hydroxytryptamine) plays a critical role in modulating gut motility, sensation and secretion and causes the diarrhea associated with carcinoid tumors. Most of the body's serotonin is produced by enterochromaffin cells, and a small amount by serotonergic neurons of the myenteric plexus. Mechanical stimuli, acidity, invading pathogens and dietary contents activate sensory receptors on enterochromaffin cells to secrete serotonin. If irritants are present in the food, the enterochromaffin cells release more serotonin to increase motility so that the noxious substance is cleared from the gut in a quick manner. This signaling has not been completely understood, but seems to involve a complex sequence of autocrine and paracrine actions. Cells respond to stimuli by releasing ATP that is converted extracellularly to ADP, which in turn acts via purinergic (P2Y) receptor-mediated Ca^{2+} signaling cascade to release high concentrations of 5-HT [114]. 5-HT then stimulates epithelial cells, intrinsic (IPANs) and extrinsic primary afferent neurons (EPANs) and part of it enters the blood stream where it is avidly taken up by platelets. Specific 5-HT receptor subtypes on different IPANs modulate the secretory reflex. Serotonin via 5HT3 receptors stimulates EPANs and results in CNS-mediated responses of nausea and discomfort. This has clinical ramifications, in that drugs that block 5HT3 receptors are very effective in controlling the nausea and vomiting, associated with chemotherapeutic treatments of cancer.

Serotonin is inactivated by being internalized via the serotonin reuptake transporter (SERT) that is present on enterocytes and neurons. The increased motility and hence diarrhea in patients with irritable bowl syndrome (IBS) has been demonstrated to be due to a reduction in SERT. Clinically, specific serotonin receptors have been (targeted) to treat diarrhea; therefore, 5-HT3 receptor antagonists such as alosetron used to treat diarrhea-predominant IBS, while tegaserod, a partial 5-HT4 agonist is used to alleviate constipation associated with IBS. However, considerable side effects preclude their use more widely in IBS [27, 34, 35].

Purine nucleotides such as adenosine also play an important role in modulating secretion in vivo. As described above in the section on serotonin, they can stimulate secretion by increasing polymorhponuclear leukocyte infiltration of the mucosa or indirectly via release of 5-HT. However, secretions evoked by mechanical stimuli are inhibited by adenosine acting via P1 purinoreceptors. Activation or inhibition of different populations of channels may be involved in these contrasting effects.

7.2.4.4 Guanylin and Nitric Oxide. As in the case of opiate peptides, studies exploring why mammalian intestine had a specific receptor for a bacterial product (heat stable enterotoxin) that stimulates Cl^- secretion, led to the discovery of the eukaryotic guanylin and uroguanylin peptides. These small endogenous peptide hormones, synthesized in goblet and columnar cells activate a membrane receptor - guanylate cyclase (GCC) to increase intracellular cGMP and elicit fluid secretion. The guanylin family of peptides regulate the gut-renal acid-base response axis and are ideal therapeutic agents targeting receptors in the intestine, kidney, and other epithelia. The neuroimmune regulator, nitric oxide, stimulates soluble guanylate cyclase to also increase cGMP. This enzyme is far more prevalent in the subepithelium of the small intestine, but is expressed in colonic epithelia. Some potential applications of the cGMP-regulating agonists include treatments for constipation as associated with IBS and sodium-sensitive hypertension.

The above is a partial description of secretagogues; additional agents that stimulate secretion and also alter intestinal motitlity are listed in Table 2.

7.3 OTHER REGULATORY EFFECTS

7.3.1 Developmental Regulation

The developing intestinal tract adapts both structurally and functionally to accommodate the varying nutritional needs at different periods in the life of a mammal. Most important of these adaptations are the ones the animal undergoes during parturition and weaning. Another contributor is the luminal microbial environment in that the newborn colon does not have as much bacterial colonization as the adult, and the time taken to reach an adult microbiota spectrum depends on a variety of factors, including the species. Other demands on the developing intestine include a maturation of its motility function, immune system, and ability to cope with the movement of large amounts of fluid. The anatomy and functions of different segments of the developing intestine exhibit specific characteristics to meet these demands. It is not surprising that studies from various laboratories, including our own, show that developmental regulation is multifactorial, species-dependent, and complex.

To maintain luminal osmolarity and adapt to the changing demands during development, the intestine also has to adjust its fluid transport properties. The ability to conserve water and salt are of greater importance in the newborn, where reserves are small and diarrhea is frequent. Compared with the adult, the neonatal small intestine exhibits greater transepithelial conductance and permeability to ions. A majority of the intestinal studies have focused on the changes in nutrient transport occurring at the critical transitions of parturition and weanling. While the adult mammalian colon is chiefly responsible for the conservation of water and electrolytes, the neonatal colon plays a role in nutrient absorption as well. The question follows whether the well characterized fluid absorptive and secretory processes of the adult, balanced by cell-specific ion transporters and defined neurohumoral regulatory cascades, are equally operative in the young animal.

Early studies on postnatal developmental changes in colonic ion transport have focused on Na^+ transport. Higher rates of Na^+ absorption as compared to the adult are seen in the rectum of human infants [10] and in the distal and proximal colon of suckling rabbits. Potter et al. [87] demonstrated that net flux of Na^+ and Cl^- was similar in the neonatal and adult rabbit colon, but in the neonate, unlike the adult, Cl^- transport is not linked to Na^+ transport. The balance of absorption and secretion are not fully geared for maximal absorption in the neonatal epithelium. This balance is shifted so that absorption predominates as the animal matures into adulthood. Features such as increased tissue permeability, decreased absorptive processes, and increased tissue receptors for enterotoxins have been suggested to contribute to the susceptibility of the neonate to diarrhea and fluid loss [115]. The picture is complicated by "cross-talk" between signaling cascades, known to exist in the adult and whether these cascades are subject to developmental changes and if so, do all cascades develop in tandem or asynchronously, the net outcome of which controls the duration and amplitude of the final response.

It is also conceivable that the developing colon may have its own "protective" mechanisms to prevent excessive fluid loss. The precise cellular basis for the susceptibility of the neonate or its ability to protect itself has not been investigated in detail. However, the relatively few ontogenic studies of ion transport have provided some interesting insights. Our laboratory investigated the ontogeny of Cl^- transport and its regulation in the rabbit colon and our evidence suggests that not only is their asynchronous "maturation" of different signaling cascades, but also mechanisms to protect excess fluid loss in the neonate. Bile acids and hydroxy fatty acids are potent secretagogues in the adult rabbit colon, and a number of pathways have been implicated, including cAMP, histamine, PGEs, $[Ca^{2+}]_i$, and PKC [22, 88]. We have demonstrated that the secondary bile acid, taurodeoxycholate (TDC), was capable of increasing $[Ca^{2+}]_i$ in the rabbit adult, but not in the neonatal and weanling colonocytes, and that this was due to a lack of PLC translocation to the membrane as well as its function (i.e., ability to generate IP3) in the neonatal and weanling animals. Our study showed the PKCδ protein is present in adult, but not in weanling or newborn colonocytes. In addition, a PKCδ inhibitor, rottlerin, and a PKCδ-specific inhibitor peptide, δV1-1, inhibited TDC-stimulated Cl^- transport in adult colonocytes [57]. Intestinal handling of bile acids is also age dependent. We found the mRNA levels of Na^+-dependent bile acid transporter (ASBT) and lipid-binding protein (LBP) are maximal in weanling and significantly lower in adult or newborn colon. We also demonstrated that weanling, but not adult colon, shows net bile acid absorption, which has relevance in enterohepatic conservation of bile acids when ileal bile acid recycling is not fully developed [123].

Of wider physiological relevance is that the age-dependency of PLC signaling is extended to the action of all Ca^{2+}-dependent secretagogues, including acetylcholine and neurotensin. However, the distal steps in calcium signaling cascade are patent in the developing colon as pharmacologically increasing $[Ca^{2+}]_i$ with the ionophore A23187-stimulated Cl^- transport in the distal colon at all ages. There are also age-related differences in the cGMP signaling cascade—adults and weanling animals, but not neonates secrete in response to the cGMP activator, heat-stable enterotoxin STa, In contrast, the cAMP-responsive system was present at an early age and we demonstrated, that in the newborn, weanling and adult rabbit distal colon forskolin and PGE1 stimulate Cl^- secretion. Thus, unlike for Ca^{2+} and cGMP signaling, the proximal steps in cAMP signaling—namely, the receptor and cyclase—were also patent in the newborn colon [21]. Thus, while the necessary transporters for Cl^- secretion are available in the neonate, the secretion is limited being triggered only by cAMP signaling and not by a variety of Ca^{2+} and cGMP-dependent neuromodulators. Taken together, these results indicate that there are protective mechanisms in the neonatal animal to prevent copious fluid loss in response to pathogens [17]. A systematic elucidation of the various signaling pathway involved in the stimulation of secretion by a number of secretagogues will provide a framework for dissecting the complexities underlying developmental changes in ion transport.

7.3.2 Systemic Effects

Intestinal transport capability depends on the metabolic status of the intestine. Thus, a well-fed gut transports more effectively, with glucose being a source of metabolic fuel throughout the length of the intestine. Even so, there is segmental variability in the preference for other metabolic fuels. For example, the small intestine uses glutamine quite effectively and short chain fatty acids, especially butyrate, are a major source of metabolic fuel in the colon.

Intestinal ion transport is altered in vivo by acid-base balance, the volume status and intestinal blood flow. Electroneutral absorption of NaCl is strongly stimulated by metabolic acidosis, and inhibited by metabolic alkalosis and this can be demonstrated in in vitro models as well. Intracellular pH and pCO_2 alter NHE activity while $[HCO_{3i}]^-$ modulate basal Cl^- secretion. In terms of volume, a decrease in intravascular volume (hemorrhage), increasing fluid absorption. Increasing sympathetic input, via cardiac baroreceptors and cardiopulmoanry mechanoreceptors, into ENS decreases secretion. In addition, alterations in plasma volume affects blood pressure and trigger the release of Angiotensin II, antidiuretic hormone or atrial natriuretic peptide; these hormones in turn may regulate intestinal fluid transport.

7.3.3 Osmotic Effects

The luminal osmolarity of the intestines, except in duodenum and proximal jejunum, is the same as that of the plasma and the intestine functions neither to dilute nor concentrate their contents. Rapid equilibration of a hypertonic lumen of duodenum and jejunum is usually accomplished by movement of water into the intestinal lumen. Luminal volume is decreased gradually by the absorptive processes and interestingly, the osmolarity of stool water is similar to plasma osmolarity.

However, a common cause of diarrhea, termed osmotic diarrhea, results from a failure to absorb a nutrient normally, resulting in the continued presence of a nonabsorbable solute within the intestinal lumen thereby increasing the osmotic load and surpassing the ability of the colon to reabsorb electrolytes. In contrast to the watery diarrhea where stool osmolarity is similar to plasma, in osmotic diarrhea it is hypertonic.

One such common nonabsorbable solute are disaccharides. Unlike amino acids which can be absorbed as di- and tri-peptides, sugars can only be absorbed as monosaccharides. The absence of a specific brush border hydrolase, that can convert disaccharides into absorbable monosaccharides, can therefore prevent accompanying fluid absorption. An example of this is lactose intolerance seen in large segments of the adult population. The absence of lactase in the adult, prevents the hydrolysis of lactose (the glucose-galactose disaccharide), thereby increasing luminal osmolarity and resulting in an increase in small bowel fluid. Osmotic diarrhea can also caused be caused by excessive sugars such as fructose and sorbitol that are present in processed foods and drinks; the intestinal capacity to absorb fructose is limited by the facilitated diffusive carrier Glut 5 and the ability of fructose to be isomerized to glucose—when this is exceeded, diarrhea ensues.

Carbohydrates and oligosaccharides that cannot be digested and absorbed in the small bowel, are converted into an absorbable solute (especially SCFA) by the action of colonic bacteria. As discussed in Section 5.2.3, SCFA are rapidly absorbed across the colonic mucosa. However, carbohydrate- induced diarrhea results if the bacterial conversion of carbohydrates to SCFAs is maximized or if the SCFA absorptive capacity of the colon is exceeded, thus allowing unmetabolized carbohydrates to pass through the colon. The partially digested carbohydrates could add to the osmotic load, exacerbating the fluid loss.

Increased intake of cations such as magnesium or anions such as sulfates, ions that are absorbed poorly by the normal gut, leads to osmotic diarrhea. This is the mechanism of action by which Mg^{2+}-based laxatives relieve constipation, and also account for the iatrogenic diarrhea observed with the intake of magnesium-containing antacids.

Osmolality is an important consideration in the design and administration of enteral nutrition to patients and in the design of ORT formulae. Reducing the osmotic load while increasing the caloric value is achieved by replacing glucose with complex carbohydrates such as rice, and using di- and tri-peptides instead of amino acid s in the formulation. This also creates a hypotonic luminal environment to enhance water absorption. These complex carbohydrates, in addition to providing sugar molecules/mosmole, are metabolized by commensal bacteria in the colon to release SCFA, which in turn promote fluid absorption [91].

7.3.4 Homocellular Regulation

Considering the extreme variability in the intestinal luminal content and osmolarity, especially in response to feeding, the enterocyte has to geared to rapidly respond to changes in the rates of ion and nutrient transport. Thus, while the properties of the two membranes of a polarized epithelial cell are distinctive, there has to be some form of "communication" and concerted action between the apical membrane and the basolateral membrane for optimal cell function. In other words, the cell needs to ensure that what enters the cell in the apical membrane, should effectively exit at the basolateral membrane at a similar rate; failure to do so in the face of a rapidly changing milieu will result in either the cell shrinking or exploding. Over 30 years ago, two phrases, "homocellular regulation" [106a] and "cross-talked" [24a] were coined to explain this communication. A good example of homocellular regulation is Na^+ absorption especially in response to a meal. The rate limiting step is the apical membrane where Na^+-glucose cotransport is maximally stimulated, but nevertheless the amount of Na^+ transiting the cell is far greater that the basal cell Na^+ content. To ensure that the intracellular environment is relatively constant there is a concomitant increase in the activity of the basolateral pump-leak processes, Na^+/K^+ pump and K^+ conductances. Similar coordination occurs during copious amounts of secretion as seen in pathological states such as watery diarrhea or in normal salivary gland physiology. A number of factors including cell volume, stretch-sensitive channels, intracellular pools of either Na^+ or Ca^{2+} and ATP, serve to regulate this critical communication and fine-tuning.

• • • •

CHAPTER 8

Intestinal Disorders and Advances Toward Better Treatment of Intestinal Disorders

The complex organization and well-honed regulation of the intestine notwithstanding, genetic, environmental, and systemic perturbations can lead to intestinal disorders. As an organ that is in direct contact with the external environment, it is not surprising that microbial pathogens are leading disrupters of intestinal function normally resulting in diarrhea. Genetic disorders could lead to either an excess of fluid secretion as seen in congenital chloridorrhea or a severe paucity of fluid secretion, resulting in meconium ileus as seen in CF. Genetic predisposition, systemic factors, and infectious agents could also lead to chronic or acute inflammation, which, in turn, can result in diarrhea. On the other hand, systemic factors, many of which remain to be elucidated, can result in constipation. This chapter provides an overview of intestinal disorders and their treatment modalities.

8.1 INFECTIONS

A variety of microbial organisms ranging from viruses to bacteria, fungi, and nematodes alter electrolyte transport, increase intestinal permeability, and trigger inflammation to result in diarrhea. They do this by a variety of mechanisms, including production of enterotoxins, disruption of tight junctions, or insertion of their own products into epithelial cells to alter host cell machinery [9, 45, 59, 85, 102].

The list of microbial pathogens is long and a few key examples are provided below.

A variety of *Shigella* species causes dysentery by producing a cytolytic toxin that damages the mucosal lining, enters the cell, inhibits protein synthesis and impairs absorption. Similarly, toxins produced by *Clostridium difficile*, *Vibrio parahaemolyticus*, *Clostridium perfringens*, and enteroadherent and enterohemorrhagic *E. coli* cause diarrhea by breaking down the cytoskeleton and regulating actin polymerization. In contrast, secretory diarrheas such as cholera and traveller's diarrhea are due to noninvasive pathogens and the enterotoxin is the agent that enters the cells to turn on the secretory machinery of the epithelium.

Cholera, most prevalent in Afganisthan, Bangladesh, India, Pakistan and other developing countries, is caused by the pathogen *Vibrio cholerae.* It produces many toxins including an enterotoxin (cholera toxin, CT), a zona occludens toxin (Zot) and a channel like protein (Figure 16). CT is an 84-kDa protein consisting of a dimeric A subunit and five identical B subunits. The A subunit contains the active toxin moiety and the B subunits binds to GM1 ganglioside abundant in the intestinal brush border membrane; this allows the A subunit to enter the enterocyte and be endocytosed. The A subunit undergoes a cleavage and the A1 peptide of this subunit contains the toxic activity. A1 acts by covalently modifying the α subunit of the stimulatory G protein Gs via adenosine diphosphate ribosylation of an arginine residue. This modification results in inhibiting the nascent GTPase activity of Gαs. The net result is that Gαs does not return to its basal state, remains activated, and triggers adeylate cyclase to produce copious amounts of cAMP. This increase in cAMP increases the activity of CFTR and inhibits the NHE3 activity to cause unregulated fluid secretion (Figure 16) Further, ZOT increases the permeability of tight junctions, which results in further fluid loss. Equally important, *Vibrio cholerae* involves components of the PINES to produce a complex response; for example in experimental models, exposure of the physically isolated small intestine to the pathogen, can elicit secretion in the colon, via involvement of the ENS. Despite these secretory effects of the toxin, the Na^+-glucose, Na^+-amino acids and SCFA absorptive pathways remain unaffected forming the basis for oral rehydration therapy (see Chapters 7 and 8).

Increase of intracellular cGMP also results in copious secretion in the intestine and strains of *E-coli* and *Yersinia enterocolitica*, produce heat-stable enterotoxins that causes traveller's diarrhea via this pathway. Similar to cAMP signaling, cGMP also activates CFTR and inhibits NHE3, but via a related yet distinct protein kinase (PKG) signaling pathway (see Section 7.1.2). *Vibrio parahaemolyticus* produces a thermostable direct haemolysin (TDH) that causes gastroenteritis and diarrhea by increasing intracellular calcium and activating the Ca^{2+}-calmodulin and protein kinase C signaling pathways. Rotavirus is the major cause of gastroenteritis among infants and young children and is transmitted by the oral-fecal route. It is a double stranded RNA virus and predominantly infects the villar enterocytes to produces a rotaviral toxin protein NSP4. This protein causes a calcium dependent increase in secretion and also inhibits disaccharidases and disrupts SGLT1 dependent fluid reabsorption, leading to watery diarrhea. The mechanism by which NSP4 causes an increase in Cl^- secretion in vivo is controversial as discussed in Section 5.2.1.5. The current thinking is NSP4 exerts transient secretory actions, has no effect on crypt cell Cl^- secretion, and under conditions of favorable electrochemical gradients, can elicit Cl^- secretion from villus cells. Despite this, there are claims that NSP4 can also stimulate Cl^- absorption in villus cells. Similarly the parasite *Cryptosporidium* causes massive watery diarrhea diarrhea in patients with cryptosporidiosis, possibly via an enterotoxin; however, the severity depends on the patient's immune status.

Multiple cell signaling molecules such as kinases and phosphatases, are recruited by many pathogens to disrupt paracellular permeability by affecting the cytoskeleton and tight junctional proteins. Enteropathogenic (EPEC) and enterohaemorrhagic *E.coli* (EHEC) are two such pathogens that decrease transepithelial resistance using different signaling cascades. The EPEC strains adhere to the intestinal cells and recruits the cell's own network of cytoskeletal machinery (Type II secretion apparatus) to inject bacterial components into the host cells. These molecules further changes actin-myosin network and alters tight junction proteins to decrease absorption and increase fluid accumulation in the intestine [9, 45]. The disruption of TJs by enteropathogenic *E. coli* is specifically associated with PKC-ζ activation [112].

Similarly, *Clostridium difficile*, an anaerobic bacteria, causes antibiotic associated pseudomembranous colitis and *C. perfringens*, responsible for food-borne illnesses, by altering intestinal permeability. The *C. difficle* toxins A and B breaks down the perijunctional acto-myosin ring, via Rho family of proteins, to increase permeability. The tight junctional claudin proteins, serve as receptors for *C. perfringens* secretes an enterotoxin (CPE) that binds to the claudin proteins in the tight junctions, thus resulting in disruption of tight junctional fibrils. Proteases that alter the junctional proteins disrupt the integrity of the tight junction and this method is used by *Bacteroides fragilis* and *Vibrio cholerae* to increase paracellular permeability leading to fluid loss.

Finally, many bacterial strains including EPEC, EHEC, Salmonella and Shigella, trigger a highly specialized cascade to stimulate ion secretion. Via an NF-KB-mediated process, these pathogens induce the expression of receptors for the peptide neurohormone, galanin, whereas uninfected cells do not possess galanin receptors. Galanin, in turn activates Cl^- secretion via Ca^{2+}-dependent signaling processes [114]. The mechanisms by which toxins act are multiple and ever-expanding and it is important to delineate these intricate processes to improve our knowledge of the pathophysiology of infectious diarrheas.

8.2 BILE ACID- AND FATTY ACID-INDUCED DIARRHEAS

Bile acids are detergent molecules derived from cholesterol and are made in the liver. The primary bile acids, cholic acid and chenodeoxycholic acid, are synthesized and conjugated with amino acids glycine and taurine in the hepatocytes, and are stored in the gall bladder. Bile acids are released via the pancreato-biliary duct in response to a meal and play a very important role in the digestion of lipids and absorption of cholesterol, fat-soluble vitamins, and lipids in the intestines. In a healthy adult, ~95% of the bile acids are reabsorbed and transported back to the liver and less than 5% reach the large intestine, where it undergoes bacterial deconjugation and dehydroxylation to form secondary bile acids such as lithocholic acid and deoxycholic acid. Secondary bile acids alter colonic function by stimulating epithelial cell secretory processes, affecting tight junctional permeability, and possibly affecting colonic motility.

Four decades ago, in vivo studies by Mekjian et al. [76] demonstrated that increases in luminal concentrations of bile acids could cause diarrhea. Thus, ileal resection, bacterial overgrowth, or other idiopathic conditions cause bile acid malabsorption and the resulting accumulation of bile acids in the colon, leads to bile acid-induced diarrhea and also impairs enterohepatic recycling. In addition to ileal resections, malabsorption of bile acids have been implicated in the diarrhea associated with intestinal inflammatory disorders such as Crohn's disease and high levels of bile acids were recently shown to play a major role in the pathophysiology of necrotizing enterocolitis [41, 93]. The bile acid-binding resin, cholestyramine, has been effectively used to treat patients with chronic diarrhea. Bile acid exposure to mucosal surfaces has been suggested to increase mucosal permeability by modulating the integrity of the tight junctions and causes mucosal damage at high concentrations (<1 mM); at higher concentrations, their effects could be dependent on a detergent action. Interestingly, the more hydrophilic bile acids such as TDC, increased permeability upon basolateral, not apical application, while lipophilic CDCA was effective from both surfaces [68]. However bile acid action may not be strictly on the paracellular pathway but could involve cellular processes.

Using cell culture, human, and animal models, we and others have studied the cellular mechanism of action of bile acids. Various second messengers have been implicated and in addition to different experimental conditions some of the observed differences could be due to inherent variabilities among the many experimental models [23, 33, 75]. The variables include concentrations examined, state of conjugation, and structure of bile acids being studied and the sidedness of exposure to bile acids. In brief, these studies have demonstrated that conjugated and unconjugated forms of dihydroxy bile acids such as CDCA and DCA induce secretion in mammalian colon, while trihydroxy bile acids such as cholic acid do not. Further, the secretory effects of bile acids have been reported to occur both at high concentration (mM range) [6, 76], and at much lower concentrations (10–100 μM) [21, 88, 116]. Structure-activity and specificity studies suggest that the Cl^- secretory effect is not due to detergent action of the bile acids. The major second messengers implicated in bile acid action are Ca^{2+}, cAMP and specific ser-threonine and tyrosine kinases [88, 116]. Thus, we have shown that the effects of bile acids on Cl^- secretion in rabbit distal colonic epithelial cells are mediated by increases in $[Ca^{2+}]_i$, PKC-δ, and tyrosine kinases; the precise sequence of these signals remain to be elucidated [57, 116]. Others have shown the involvement of cAMP in intact epithelial preparations, using higher concentrations of bile acids and implied that bile acids could be causing their effect by stimulating the release of histamine and prostaglandins from subepithelial cells [33]. Interestingly, the secretory effects of all Ca^{2+}-dependent secretagogues, including bile acids are developmentally regulated and manifest only in the adult rabbit colon [21, 88, 116].

These studies on cellular actions of bile acids imply that rather than as a transported "solute" they may be acting as an "extracellular signal" to activate specific "receptors." It is only in the last two decades, that this additional and important role of bile acids as signaling molecules has been identified. In 1999, three independent studies reported the presence of a nuclear farnesoid X receptor, capable of binding bile acids and suggested that much like steroid hormones and vitamin D, bile acids act as ligands to activate various signaling cascades. Within the past decade a membrane receptor for bile acids termed, TGR5 has also been identified in the liver and gastrointestinal tract. This receptor belongs to the superfamily of HHMS receptors and is coupled to Gαs, and acts via the cAMP pathway. The relevance of either or both of these receptors in bile acid stimulated diarrhea remains to be elucidated.

Another consequence of fat malabsorption is the increase in long chain fatty acids in the colon; these are generated by the action of triglyceride lipases in the small intestine but due to poor absorption, reach the colon. Colonic bacterial metabolism hydroxylates these long-chain fatty acids and the hydroxylated derivatives are potent secretagogues. For example, oral castor oil, is relatively nontoxic, but upon being hydroxylated to ricinoleic acid, it becomes a potent secretagogue and is used as a laxative. Therefore excess long chain fatty acids in the colonic lumen in conditions such as sprue, will result in fluid secretion. Specific fatty acid transporters have been identified in the intestine and their mechanisms of action could involve a combination of processes including indirect stimulation via release of prostaglandins and/or triggering intracellular cascades [1, 39].

8.3 INFLAMMATION

A full treatment of diarrheas associated with inflammation is beyond the scope of this book, and therefore a very brief overview is provided here. The clinical correlation between intestinal inflammation and diarrhea is obvious (Chapter 2), with ulceration, exudation of protein, changes in motility, and loss of absorptive surface area implicated as causing the fluid losses of inflammatory diarrheas. Intestinal inflammation increases the number of immunocytes, the cause of the inflammation determining the type of inflammatory cells, e.g., acute bacterial infections increase polymorphonuclear leukocytes, whereas parasitic infections dramatically enlarge the mast cell population, and celiac sprue is characterized by intraepithelial lymphocytes. In inflammatory bowel diseases, both Crohn's disease and in ulcerative colitis, there is activation of all components of GALT with an increase in IgG-secreting cells [70, 97]. Thus, the cause of the inflammatory reaction may determine the type of immunocytes recruited, the range of cytokines released, and the specific effects on transport and motility. Therefore, a one-size-fits-all modality cannot be used for treating IBDs. As discussed in Chapter 7, many inflammatory mediators are potent secretagogues, including peptides, eicosanoids, and oxidants. These mediators either interact directly with epithelia to alter ion transport and barrier function or elicit these effects indirectly by activating other PINES elements (Figure 15). Thus, prostaglandins are a prime example of an immunomodulator central to the secretory response associated with inflammation.

The etiology of inflammatory bowel diseases are complex and genetic and autoimmune components have been implicated. Another similar disease with complex etiology is celiac disease, also known as celiac sprue. This disorder can affect children and adults and is characterized by gluten intolerance that results in inflammation and atrophy of the mucosa of the small intestine. It is among the most common lifelong disorders worldwide, and it affects 0.5% to 1% of the general population in the United States and other developed countries. Recent epidemiologic studies showed that the prevalence of this disease is higher than estimated in other parts of the world including Asia [14]. The first signal of this disease is the appearance of serum antibodies. The common symptoms are chronic diarrhea, abdominal pain, etc. Affected children may have impaired growth. The clinical manifestation can also be "atypical" and patients may suffer from anemia, osteoporosis, etc. Serological screen may also detect some cases without any symptoms.

The genetics of celiac disease is complex. In addition to histocompatibility leukocyte antigen (HLA) class II genes HLA-DQ2 and HLA-DQ8, many other genes are involved in the development of the disease. Once the diagnosis is established, a gluten-free diet is strictly recommended; however this in of itself may attenuate but not obliterate the symptoms. With the recent studies on cellular and molecular mechanism of celiac disease, there are several potential new therapies. In 2007, a double-blind, randomized clinical trial showed that Larazotide acetate (AT-1001) prevented gluten-caused increase of intestinal permeability compared with placebo in celiac patients [84]. The second non dietary treatment is enzyme therapy. Clinical trials are going on to test the gluten detoxifying effects of some endopeptidases, including ALV003 and AN-PE [62]. The chronic diarrhea associated with this disease is due to the reduction in absorptive surface area, thereby letting secretion predominate.

8.4 GENETIC DISEASES (CYSTIC FIBROSIS AND CONGENITAL CHLORIDORRHEA)

Cystic fibrosis is an inherited autosomal recessive disease that affects multiple organ systems including the lung, sweat glands, pancreas, testis, and intestine; the disease derives its name from pancreatic obstructions (also see Section 5.2.1). The major causes of mortality are respiratory disease and pancreatic insufficiency arising out of the basic CF defect. While in the early 1950's, life expectancy was below 5 years of age, significant understanding of the basis of this disease has led to improved management, and the life expectancy is currently well over 35 years of age (see Section 5.2.1). Mutations in the CF gene, results in a defective protein CFTR, a key Cl channel protein in exocrine tissues (Figure 10). Impaired or absent CFTR function results in the elaboration of thick, sticky mucus-enriched secretions that affect all exocrine gland secretions and in particular in the lungs and pancreas. A common pathology in the CF infant is meconium ileus and ≈ 15% of CF adults present with distal intestinal obstructive syndrome. Unlike humans, mice have a strong alternative Cl^- secretory "salvage" pathway in the lung. The defect in the CF mouse is manifested as meconium ileus and, unless treated with an osmotic laxative, results in early mortality of this useful animal model.

The different ways by which the mutations in CFTR alter function are discussed in Section 5.2; these include defects in protein production, processing, defective channel regulation and conduction. Although identification of the CF gene engendered great hope in developing a gene therapy strategy for curing the disease, the pluripotent nature of CFTR and its complex phenotype have made gene therapy a difficult approach. Amongst other challenges two major problems need to be resolved before the application of gene therapy to CF patients. First, a perfect gene delivery system needs to be found. Second, the frequency and the length of treatment need to be determined (NIH). Therefore other therapeutic strategies are being pursued. In addition to nutrition supplements to tackle the pancreatic malfunction, and the use of amiloride + antibiotics + mechanical dissolution to relieve respiratory distress, pharmacological and combinatorial chemistry approaches are proving to be of use.

The most common type of mutation in CF patients is deltaF508, which disrupts the processing and trafficking of CFTR, and only a small amount of mutated CFTR makes it to the apical membrane (see Chapter 5.2). This protein forms Cl^- channels with normal conduction but defective gating. One current approach to remedy this situation is to use small molecules, called CFTR correctors, that overcome the processing defect of delta F508 and another named, CFTR potentiators, which repair the gating defect of the mutated CFTR. Some of the correctors that are currently used are quinazolinone VRT-325, thapsigargin and curcumin; they act by "tricking" the endoplasmic reticular quality control system to recognize the mutant CFTR as "normal" rather than defective. This will then, allow the CFTR to traffic to the plasma membrane and restore functional Cl^- channel activity rather than have it be sent to the proteasomes for degradation. CFTR potentiators include 8-cyclopentyl-1, 3-dipropylxanthine (CPX), 1,3-diallyl-8-cyclohexylxanthine (DAX) and genistein, which act by inhibiting a phosphodiesterase or directly binding to NBD of CFTR and increasing Cl^- secretion. Protoesome inhibitors have also been used to decrease degradation of the mutated protein and in combination with the CFTR correctors and potentiators the functional activity of CFTR can be restored. Another approach to help increase Cl^- secretion and hence fluid is to activate other Cl^- channels that are present in the apical membrane, including TMEM16A (see Chapter 5). The identity of TMEM16A activators are still unknown and could be a valuable discovery leading to the development of CFTR by pass therapy. However, the efficacy of such a drug in the intestine will depend on whether nonCFTR Cl^- channels, including TMEM16A, can be unequivocally demonstrated in the apical membrane of enterocytes.

Congenital chloride diarrhea is a rare autosomal-recessive mutation on chromosome 7 that leads to a defect in brush border DRA (SLC26A3), a Cl^-/HCO_3^- exchanger. This results in decreased fluid absorption in the intestine leading to diarrhea. This genetic defect, although found world-wide, is most prevalent in Finland. In a fetal stage, abnormal fluid absorption begins inside the uterus and causes maternal polyhydramnios with excess fluid in the amniotic sac. Immediately after parturition, this defect leads to dehydration, increased bilirubin levels, abdmoninal distension and watery diarrhea. However the nutrient absorption in these patients remains unaffected. The pH of their stool is acidic, and they develop metabolic alkalosis. In fact this is used to diagnose this congenital defect as all other forms of diarrhea leads to metabolic acidosis. Aggressive replacement of NaCl and KCl and use of butyrate as the energy source has been effective to treat symptoms and results in normal growth and development of the infant.

8.5 CONSTIPATION

Constipation is one of the most common gastrointestinal complaints in the United States, and more than 4 million Americans have frequent constipation (NIH). Constipation is defined medically as fewer than three stools per week and severe constipation as less than one stool per week. Constipation is a symptom, not a disease, and may be idiopathic or secondary to other diseases. Severe constipation may greatly affect patients' quality of life. According to Suares and Ford [111], the pooled prevalence of chronic idiopathic constipation in all studies was 14%, higher in women, and increases with age and lower socioeconomic status. The prevalence was markedly higher in subjects who also reported irritable bowel syndrome [111].

Many of the common causes for constipation are not having enough fiber in the diet, not enough liquid consumed, lack of physical activity, side effects of medications, and other diseases. Neurological disorders, metabolic and endocrine dysfunction, and autoimmune diseases such as lupus can cause constipation as well. When neurons that stimulate secretion from the crypts are less active, the fluidity of luminal contents in the intestine is decreased and results in constipation. Increased sympathetic activity and norepinephrine binding to α-2a receptors in enteric neurons could also lead to constipation. Further, mast cells secrete proteases that forms angiotensin II, which results in the release of sympathetic postganglionic neurotransmitter norepinephrine via activation of AT1 receptor. This suppresses intestinal blood flow and causes the intestine to remain in a state of absorption, thus promoting constipation [125].

Colonic epithelium has both absorptive and secretive functions, and the various underlying transporters (Chapter 5) work in coordination to ensure fluid and ion balance. Defective net ion secretion or increased net absorption leads to decreased water secretion, hence constipation. Amitiza (lubiprostone) is used to treat chronic idiopathic constipation since it can increase Cl^- secretion. It has been shown to activate ClC2 Cl^- channels in a number of model systems, including T84 cells [19]. However, we [4] and Bijvelds et al. [10] showed that lubiprostone stimulated intestinal Cl^- secretion via CFTR, not ClC2, in T84 cells. In the intestinal epithelium of wild-type and CF mice and of normal and CF patients, Bijvelds et al. [10] confirmed CFTR's involvement. Although there is the controversy over how the drug acts, it effectively increases intestinal secretion and functions to relieve functional constipation. Its potential as a therapeutic strategy to bypass the Cl^- defect in CF patients remains to be established.

8.6 ORAL REHYDRATION THERAPY

No discussion of therapeutic strategies is complete without a mention of the medical marvel of the 20th century—the introduction and wide-spread implementation of the use of oral rehydration therapy. This treatment alone reduced mortality due to diarrhea in children under the age of 5, from >5 million in 1978 to 1.3 million in 2002. The essence of this remedy exists in traditional remedies in various cultures around the globe but the past 50 years have helped us unravel the scientific basis for its success and streamline its formulation and use. The beauty of this treatment is its simplicity of administration and low cost. This simple therapy depends on timely replenishment of salts and sugars and therefore fluid; this intervention prevents dehydration, the major cause of mortality in infectious diarrheas. There are three features which are integral to the success of oral rehydration therapy: a. The observation that active glucose transport in the small intestine and that of short chain fatty acids in the colon are not affected by enterotoxins (secretagogues) that increase cAMP or cGMP; b. The presence of commensal bacteria in the colon that can digested complex carbohydrates to release SCFA; and c. The presence of SCFA transport mechanisms in the colon where SCFA serve to drive Na^+ and therefore fluid absorption as well as being another important source of metabolic fuel [91]. However, even this formulation cannot be applied readily to all forms of diarrhea and improved formulations tailor-made for different diarrheal disorders are needed. Key to continued success are increased education, improved sanitation and that any formulation needs to be readily available and inexpensive. It is therefore encouraging to note that the NIDDK as well as philanthropic organizations such as the Bill and Melinda Gates Foundation have marked pharmacotherapy for diarrheal diseases as a top priority.

• • • •

CHAPTER 9
Conclusion

Intestinal disorders resulting in diarrhea are multifactorial, complex and remain to be a major threat to human health. Infectious diarrheal diseases continue to pose a major threat in the world's most densely populated areas. A better understanding of the molecular underpinnings of intestinal electrolyte and fluid transport in health and disease, will allow us to develop more accurate therapeutic strategies.

• • • •

References

[1] Abumrad NA. CD36 may determine our desire for dietary fats. *Journal of Clinical Investigation* 115: 2965–7, 2005.

[2] Alberts B, Johnson S, Lewis J, Roff M, Roberts K, and Walter P. *Molecular Biology of the Cell.* New York: Garland Science, Taylor & Francis Group, 2007.

[3] Anderson JM, and Van Itallie CM. Tight junctions and the molecular basis for regulation of paracellular permeability. *American Journal of Physiology* 269: G467–75, 1995.

[4] Ao M, Venkatasubramanian J, Boonkaewwan C, Ganesan N, Syed A, Benya RV, and Rao MC. Lubiprostone activates Cl– secretion via cAMP signaling and increases membrane CFTR in the human colon carcinoma cell line, T84. *Digestive Diseases and Sciences* 56: 339–51, 2011.

[5] Bachmann O, Riederer B, Rossmann H, Groos S, Schultheis PJ, Shull GE, Gregor M, Manns MP, and Seidler U. The Na^+/H^+ exchanger isoform 2 is the predominant NHE isoform in murine colonic crypts and its lack causes NHE3 upregulation. *American Journal of Physiology* 287: G125–33, 2004.

[6] Bachmann O, Juric M, Seidler U, Manns MP and Yu H. Basolateral ion transporters involved in colonic epithelial electrolyte absorption, anion secretion and cellular homeostasis. *Acta Physiologica* 201: 33–46, 2011.

[7] Barrett KE. New ways of thinking about (and teaching about) intestinal epithelial function. *Advances in Physiology Education* 32: 25–34, 2008.

[8] Benga G, Popescu O, Borza V, Pop VI, Muresan A, Mocsy I, Brain A, and Wrigglesworth JM. Water permeability in human erythrocytes: identification of membrane proteins involved in water transport. *European Journal of Cell Biology* 41: 252–62, 1986.

[9] Berkes J, Viswanathan VK, Savkovic SD, and Hecht G. Intestinal epithelial responses to enteric pathogens: effects on the tight junction barrier, ion transport, and inflammation. *Gut* 52: 439–51, 2003.

[10] Bijvelds MJ, Bot AG, Escher JC, and De Jonge HR. Activation of intestinal Cl– secretion by lubiprostone requires the cystic fibrosis transmembrane conductance regulator. *Gastroenterology* 137: 976–85, 2009.

[11] Binder HJ, Rajendran V, Sadasivan V, and Geibel JP. Bicarbonate secretion: a neglected aspect of colonic ion transport. *Journal of Clinical Gastroenterology* 39: S53–8, 2005.

[12] Bjerknes M, and Cheng H. Gastrointestinal stem cells. II. Intestinal stem cells. *American Journal of Physiology* 289: G381–7, 2005.

[13] Bookstein C, DePaoli AM, Xie Y, Niu P, Musch MW, Rao MC, and Chang EB. Na+/H+ exchangers, NHE-1 and NHE-3, of rat intestine. Expression and localization. *Journal of Clinical Investigation* 93: 106–13, 1994.

[14] Catassi C, and Fasano A. Celiac disease. *Current Opinion in Gastroenterology* 24: 687–91, 2008.

[15] Cereijido M, Contreras RG, Flores-Benitez D, Flores-Maldonado C, Larre I, Ruiz A, and Shoshani L. New diseases derived or associated with the tight junction. *Archives of Medical Research* 38: 465–78, 2007.

[16] Cereijido M, Contreras RG, Shoshani L, Flores-Benitez D, and Larre I. Tight junction and polarity interaction in the transporting epithelial phenotype. *Biochimica et Biophysica Acta* 1778: 770–93, 2008.

[17] Chang E, and Rao M. Intestinal mediators of intestinal electrolyte transport. In: *Diarrheal Diseases*, edited by Field M. New York: Elsevier, 1991, pp. 49–72.

[18] Chang EB, Bergenstal RM, and Field M. Diarrhea in streptozocin-treated rats. Loss of adrenergic regulation of intestinal fluid and electrolyte transport. *Journal of Clinical Investigation* 75: 1666–70, 1985.

[19] Cuppoletti J, Malinowska DH, Tewari KP, Li Q J, Sherry AM, Patchen ML, and Ueno R. SPI-0211 activates T84 cell chloride transport and recombinant human ClC-2 chloride currents. *Am J Physiol Cell Physiol* 287: C1173–83, 2004.

[20] Denker BM, Smith BL, Kuhajda FP, and Agre P. Identification, purification, and partial characterization of a novel Mr 28,000 integral membrane protein from erythrocytes and renal tubules. *Journal of Biological Chemistry* 263: 15634–42, 1988.

[21] Desai GN, Sahi J, Reddy PM, Venkatasubramanian J, Vidyasagar D, and Rao MC. Chloride transport in primary cultures of rabbit colonocytes at different stages of development. *Gastroenterology* 111: 1541–50, 1996.

[22] Devor DC, Sekar MC, Frizzell RA, and Duffey ME. Taurodeoxycholate activates potassium and chloride conductances via an IP3-mediated release of calcium from intracellular stores in a colonic cell line (T84). *Journal of Clinical Investigation* 92: 2173–81, 1993.

[23] Dharmsathaphorn K, Huott PA, Vongkovit P, Beuerlein G, Pandol SJ, and Ammon HV. Cl^- secretion induced by bile salts. A study of the mechanism of action based on a cultured colonic epithelial cell line. *Journal of Clinical Investigation* 84: 945–53, 1989.

[24] Diamond JM, and Bossert WH. Standing-gradient osmotic flow. A mechanism for coupling of water and solute transport in epithelia. *Journal of General Physiology* 50: 2061–83, 1967.

[24a] Diamond JM. Transcellular cross-talk between epithelial cell membranes. *Nature*. 300: 683–5, 1982.

[25] Dudeja PK, Rao DD, Syed I, Joshi V, Dahdal RY, Gardner C, Risk MC, Schmidt L, Bavishi D, Kim KE, Harig JM, Goldstein JL, Layden TJ, and Ramaswamy K. Intestinal distribution of human Na^+/H^+ exchanger isoforms NHE-1, NHE-2, and NHE-3 mRNA. *American Journal of Physiology* 271: G483–93, 1996.

[26] Faham S, Watanabe A, Besserer GM, Cascio D, Specht A, Hirayama BA, Wright EM, and Abramson J. The crystal structure of a sodium galactose transporter reveals mechanistic insights into Na+/sugar symport. *Science (New York, NY)* 321: 810–4, 2008.

[27] Farthing MJ. Antisecretory drugs for diarrheal disease. *Digestive Diseases (Basel, Switzerland)* 24: 47–58, 2006.

[28] Field M. Intestinal ion transport and the pathophysiology of diarrhea. *Journal of Clinical Investigation* 111: 931–43, 2003.

[29] Flores CA, Cid LP, Sepulveda FV, and Niemeyer MI. TMEM16 proteins: the long awaited calcium-activated chloride channels? *Brazilian Journal of Medical and Biological Research = Revista brasileira de pesquisas medicas e biologicas / Sociedade Brasileira de Biofisica [et al.]* 42: 993–1001, 2009.

[30] Flores CA, Melvin JE, Figueroa CD, and Sepulveda FV. Abolition of Ca2+-mediated intestinal anion secretion and increased stool dehydration in mice lacking the intermediate conductance Ca2+-dependent K+ channel Kcnn4. *Journal of Physiology* 583: 705–17, 2007.

[31] Ganapathy V, Thangaraju M, Gopal E, Martin PM, Itagaki S, Miyauchi S, and Prasad PD. Sodium-coupled monocarboxylate transporters in normal tissues and in cancer. *AAPS Journal* 10: 193–9, 2008.

[32] Geibel JP. Secretion and absorption by colonic crypts. *Annual Review of Physiology* 67: 471–90, 2005.

[33] Gelbmann CM, Schteingart CD, Thompson SM, Hofmann AF, and Barrett KE. Mast cells and histamine contribute to bile acid-stimulated secretion in the mouse colon. *Journal of Clinical Investigation* 95: 2831–9, 1995.

[34] Gershon MD, and Liu MT. Serotonin and neuroprotection in functional bowel disorders. *Neurogastroenterol Motil* 19 Suppl 2: 19–24, 2007.

[35] Gershon MD, and Tack J. The serotonin signaling system: from basic understanding to drug development for functional GI disorders. *Gastroenterology* 132: 397–414, 2007.

[36] Giepmans BN. Gap junctions and connexin-interacting proteins. *Cardiovascular Research* 62: 233–45, 2004.

[37] Greger R. Role of CFTR in the colon. *Annual Review of Physiology* 62: 467–91, 2000.

[38] Guan Y, Dong J, Tackett L, Meyer JW, Shull GE, and Montrose MH. NHE2 is the main apical NHE in mouse colonic crypts but an alternative Na+-dependent acid extrusion mechanism is upregulated in NHE2-null mice. *American Journal of Physiology* 291: G689–99, 2006.

[39] Hajri T, and Abumrad NA. Fatty acid transport across membranes: relevance to nutrition and metabolic pathology. *Annual Review of Nutrition* 22: 383–415, 2002.

[40] Halestrap AP, and Meredith D. The SLC16 gene family-from monocarboxylate transporters (MCTs) to aromatic amino acid transporters and beyond. *Pflugers Archiv* 447: 619–28, 2004.

[41] Halpern MD, Holubec H, Saunders TA, Dvorak K, Clark JA, Doelle SM, Ballatori N, and Dvorak B. Bile acids induce ileal damage during experimental necrotizing enterocolitis. *Gastroenterology* 130: 359–72, 2006.

[42] Hansen MB, and Witte AB. The role of serotonin in intestinal luminal sensing and secretion. *Acta Physiologica (Oxford, England)* 193: 311–23, 2008.

[43] Hatch M, and Freel RW. Electrolyte transport across the rabbit caecum in vitro. *Pflugers Archiv* 411: 333–8, 1988.

[44] He P, and Yun CC. Mechanisms of the regulation of the intestinal Na^+/H^+ exchanger NHE3. *Journal of Biomedicine and Biotechnology* 2010: 238080, 2010.

[45] Hecht G. Microbes and microbial toxins: paradigms for microbial-mucosal interactions. VII. Enteropathogenic Escherichia coli: physiological alterations from an extracellular position. *American Journal of Physiology* 281: G1–7, 2001.

[46] Hediger MA, and Rhoads DB. Molecular physiology of sodium-glucose cotransporters. *Physiological Reviews* 74: 993–1026, 1994.

[47] Heitzmann D, and Warth R. Physiology and pathophysiology of potassium channels in gastrointestinal epithelia. *Physiological Reviews* 88: 1119–82, 2008.

[48] Helliwell PA, Richardson M, Affleck J, and Kellett GL. Regulation of GLUT5, GLUT2 and intestinal brush-border fructose absorption by the extracellular signal-regulated kinase, p38 mitogen-activated kinase and phosphatidylinositol 3-kinase intracellular signalling pathways: implications for adaptation to diabetes. *Biochemical Journal* 350 Pt 1: 163–9, 2000.

[49] Helliwell PA, Richardson M, Affleck J, and Kellett GL. Stimulation of fructose transport across the intestinal brush-border membrane by PMA is mediated by GLUT2 and dynamically regulated by protein kinase C. *Biochemical Journal* 350 Pt 1: 149–54, 2000.

[50] Hoglund P, Haila S, Socha J, Tomaszewski L, Saarialho-Kere U, Karjalainen-Lindsberg ML, Airola K, Holmberg C, de la Chapelle A, and Kere J. Mutations of the Down-regulated in adenoma (DRA) gene cause congenital chloride diarrhoea. *Nature Genetics* 14: 316–9, 1996.

[51] Hoogerwerf WA, Tsao SC, Devuyst O, Levine SA, Yun CH, Yip JW, Cohen ME, Wilson PD, Lazenby AJ, Tse CM, and Donowitz M. NHE2 and NHE3 are human and rabbit intestinal brush-border proteins. *American Journal of Physiology* 270: G29–41, 1996.

[52] Horisberger JD, and Chraibi A. Epithelial sodium channel: a ligand-gated channel? *Nephron* 96: p37–41, 2004.

[53] Ikarashi N, Mochiduki T, Takasaki A, Ushiki T, Baba K, Ishii M, Kudo T, Ito K, Toda T, Ochiai W, and Sugiyama K. A mechanism by which the osmotic laxative magnesium sulphate increases the intestinal aquaporin 3 expression in HT-29 cells. *Life Sciences* 88: 194–200, 2011.

[54] Ikarashi N, Ushiki T, Mochizuki T, Toda T, Kudo T, Baba K, Ishii M, Ito K, Ochiai W, and Sugiyama K. Effects of magnesium sulphate administration on aquaporin 3 in rat gastrointestinal tract. *Biological and Pharmaceutical Bulletin* 34: 238–42, 2011.

[55] Jacob P, Rossmann H, Lamprecht G, Kretz A, Neff C, Lin-Wu E, Gregor M, Groneberg DA, Kere J, and Seidler U. Down-regulated in adenoma mediates apical Cl–/HCO3-exchange in rabbit, rat, and human duodenum. *Gastroenterology* 122: 709–24, 2002.

[56] Jentsch TJ, Friedrich T, Schriever A, and Yamada H. The CLC chloride channel family. *Pflugers Archiv* 437: 783–95, 1999.

[57] Kanchanapoo J, Ao M, Prasad R, Moore C, Kay C, Piyachaturawat P, and Rao MC. Role of protein kinase C-delta in the age-dependent secretagogue action of bile acids in mammalian colon. *Am J Physiol Cell Physiol* 293: C1851–61, 2007.

[58] Kiela PR, Xu H, and Ghishan FK. Apical Na^+/H^+ exchangers in the mammalian gastrointestinal tract. *J Physiol Pharmacol* 57 Suppl 7: 51–79, 2006.

[59] Kunzelmann K, and McMorran B. First encounter: how pathogens compromise epithelial transport. *Physiology (Bethesda)* 19: 240–4, 2004.

[60] Kwon O, Corrigan G, Myers BD, Sibley R, Scandling JD, Dafoe D, Alfrey E, and Nelson WJ. Sodium reabsorption and distribution of Na+/K+-ATPase during postischemic injury to the renal allograft. *Kidney International* 55: 963–75, 1999.

[61] Laforenza U, Gastaldi G, Polimeni M, Tritto S, Tosco M, Ventura U, Scaffino MF, and Yasui M. Aquaporin-6 is expressed along the rat gastrointestinal tract and upregulated by feeding in the small intestine. *BMC Physiology* 9: 18, 2009.

[62] Crespo Pérez L, Castillejo de Villasante G, Cano Ruiz A, and León F. Non-dietary therapeutic clinical trials in coeliac disease. *European Journal of Internal Medicine* 23: 9–14, 2012.

[63] Li C, and Naren AP. CFTR chloride channel in the apical compartments: spatiotemporal coupling to its interacting partners. *Integr Biol (Camb)* 2: 161–77, 2010.

[64] Lipkin M. Growth and development of gastrointestinal cells. *Annual Review of Physiology* 47: 175–97, 1985.

[65] Lissner S, Nold L, Hsieh CJ, Turner JR, Gregor M, Graeve L, and Lamprecht G. Activity and PI3-kinase dependent trafficking of the intestinal anion exchanger downregulated in adenoma depend on its PDZ interaction and on lipid rafts. *American Journal of Physiology* 299: G907–20, 2010.

[66] Loffing J, and Schild L. Functional domains of the epithelial sodium channel. *J Am Soc Nephrol* 16: 3175–81, 2005.

[67] Loo DD, Zeuthen T, Chandy G, and Wright EM. Cotransport of water by the Na+/glucose cotransporter. *Proceedings of the National Academy of Sciences of the United States of America* 93: 13367–70, 1996.

[68] Lowes S, and Simmons NL. Human intestinal cell monolayers are preferentially sensitive to disruption of barrier function from basolateral exposure to cholic acid: correlation with membrane transport and transepithelial secretion. *Pflugers Archiv* 443: 265–73, 2001.

[69] Ma T, Song Y, Gillespie A, Carlson EJ, Epstein CJ, and Verkman AS. Defective secretion of saliva in transgenic mice lacking aquaporin-5 water channels. *Journal of Biological Chemistry* 274: 20071–4, 1999.

[70] MacDermott RP. Alterations in the mucosal immune system in ulcerative colitis and Crohn's disease. *Med Clin North Am* 78: 1207–31, 1994.

[71] Malakooti J, Saksena S, Gill RK, and Dudeja PK. Transcriptional regulation of the intestinal luminal Na and Cl transporters. *Biochemical Journal* 435: 313–25, 2011.

[72] Mandel LJ, Bacallao R, and Zampighi G. Uncoupling of the molecular 'fence' and paracellular 'gate' functions in epithelial tight junctions. *Nature* 361: 552–5, 1993.

[73] Marchiando AM, Graham WV, and Turner JR. Epithelial barriers in homeostasis and disease. *Annual Review of Pathology* 5: 119–44, 2010.

[74] Matos JE, Sausbier M, Beranek G, Sausbier U, Ruth P, and Leipziger J. Role of cholinergic-activated KCa1.1 (BK), KCa3.1 (SK4) and KV7.1 (KCNQ1) channels in mouse colonic Cl^- secretion. *Acta Physiologica (Oxford, England)* 189: 251–8, 2007.

[75] Mauricio AC, Slawik M, Heitzmann D, von Hahn T, Warth R, Bleich M, and Greger R. Deoxycholic acid (DOC) affects the transport properties of distal colon. *Pflugers Archiv* 439: 532–40, 2000.

[76] Mekjian HS, Phillips SF, and Hofmann AF. Colonic secretion of water and electrolytes induced by bile acids: perfusion studies in man. *Journal of Clinical Investigation* 50: 1569–77, 1971.

[77] Morris AP, Scott JK, Ball JM, Zeng CQ, O'Neal WK, and Estes MK. NSP4 elicits age-dependent diarrhea and Ca(2+) mediated I(−) influx into intestinal crypts of CF mice. *American Journal of Physiology* 277: G431–44, 1999.

[78] Morth JP, Pedersen BP, Buch-Pedersen MJ, Andersen JP, Vilsen B, Palmgren MG, and Nissen P. A structural overview of the plasma membrane Na^+,K^+-ATPase and H^+-ATPase ion pumps. *Nature Reviews* 12: 60–70, 2011.

[79] Muller JM, Debaigt C, Goursaud S, Montoni A, Pineau N, Meunier AC, and Janet T. Unconventional binding sites and receptors for VIP and related peptides PACAP and PHI/PHM: an update. *Peptides* 28: 1655–66, 2007.

[80] Murek M, Kopic S, and Geibel J. Evidence for intestinal chloride secretion. *Exp Physiol* 95: 471–85, 2010.

[81] Musch MW, Arvans DL, Wu GD, and Chang EB. Functional coupling of the downregulated in adenoma Cl–/base exchanger DRA and the apical Na^+/H^+ exchangers NHE2 and NHE3. *American Journal of Physiology* 296: G202–10, 2009.

[82] Nelson WJ. Regulation of cell surface polarity from bacteria to mammals. *Science (New York, NY)* 258: 948–55, 1992.

[83] Palmer LG. Ion selectivity of epithelial Na channels. *Journal of Membrane Biology* 96: 97–106, 1987.

[84] Paterson BM, Lammers KM, Arrieta MC, Fasano A, and Meddings JB. The safety, tolerance, pharmacokinetic and pharmacodynamic effects of single doses of AT-1001 in coeliac disease subjects: a proof of concept study. *Alimentary Pharmacology and Therapeutics* 26: 757–66, 2007.

[85] Petri WA, Jr., Miller M, Binder HJ, Levine MM, Dillingham R, and Guerrant RL. Enteric infections, diarrhea, and their impact on function and development. *Journal of Clinical Investigation* 118: 1277–90, 2008.

[86] Pinto D, and Clevers H. Wnt control of stem cells and differentiation in the intestinal epithelium. *Experimental Cell Research* 306: 357–63, 2005.

[87] Potter GD, and Burlingame SM. Ion transport by neonatal rabbit distal colon. *American Journal of Physiology* 250: G754–9, 1986.

[88] Potter GD, Sellin JH, and Burlingame SM. Bile acid stimulation of cyclic AMP and ion transport in developing rabbit colon. *Journal of Pediatric Gastroenterology and Nutrition* 13: 335–41, 1991.

[89] Powell DW, Adegboyega PA, Di Mari JF, and Mifflin RC. Epithelial cells and their neighbors I. Role of intestinal myofibroblasts in development, repair, and cancer. *American Journal of Physiology* 289: G2–7, 2005.

[90] Rao M. Absorption and secretion of water and electrolytes. In: *Small Bowel Disorders*, edited by Ratnaike R. London: Hodder Headline Group, 2000, pp. 116–34.

[91] Rao MC. Oral rehydration therapy: new explanations for an old remedy. *Annual Review of Physiology* 66: 385–417, 2004.

[92] Riordan JR. CFTR function and prospects for therapy. *Annual Review of Biochemistry* 77: 701–26, 2008.

[93] Robb BW, and Matthews JB. Bile salt diarrhea. *Current Gastroenterology Reports* 7: 379–83, 2005.

[94] Rowe SM, Miller S, and Sorscher EJ. Cystic fibrosis. *New England Journal of Medicine* 352: 1992–2001, 2005.

[95] Russo MA, Hogenauer C, Coates SW, Jr., Santa Ana CA, Porter JL, Rosenblatt RL, Emmett M, and Fordtran JS. Abnormal passive chloride absorption in cystic fibrosis jejunum functionally opposes the classic chloride secretory defect. *Journal of Clinical Investigation* 112: 118–25, 2003.

[96] Saksena S, Tyagi S, Goyal S, Gill RK, Alrefai WA, Ramaswamy K, and Dudeja PK. Stimulation of apical Cl/HCO(OH) exchanger, SLC26A3 by neuropeptide Y is lipid raft dependent. *American Journal of Physiology* 299: G1334–43, 2010.

[97] Sandle GI. Pathogenesis of diarrhea in ulcerative colitis: new views on an old problem. *J Clin Gastroenterol* 39: S49–52, 2005.

[98] Sandle GI, and Hunter M. Apical potassium (BK) channels and enhanced potassium secretion in human colon. *QJM* 103: 85–9, 2010.

[99] Sangan P, Brill SR, Sangan S, Forbush B, 3rd, and Binder HJ. Basolateral K-Cl cotransporter regulates colonic potassium absorption in potassium depletion. *Journal of Biological Chemistry* 275: 30813–6, 2000.

[100] Sangan P, Rajendran VM, Geibel JP, and Binder HJ. Cloning and expression of a chloride-dependent Na+-H+ exchanger. *Journal of Biological Chemistry* 277: 9668–75, 2002.

[101] Sartor R, and Powell D. Mechanisms of diarrhea in intestinal inflammation and hypersensitivity. In: *Diarrheal Diseases*, edited by Field M. NY: Elsevier, 1991, pp. 75–114.

[102] Sasaki M, Sitaraman SV, Babbin BA, Gerner-Smidt P, Ribot EM, Garrett N, Alpern JA, Akyildiz A, Theiss AL, Nusrat A, and Klapproth JM. Invasive Escherichia coli are a feature of Crohn's disease. *Laboratory Investigation; A journal of Technical Methods and Pathology* 87: 1042–54, 2007.

[103] Sausbier M, Matos JE, Sausbier U, Beranek G, Arntz C, Neuhuber W, Ruth P, and Leipziger J. Distal colonic K(+) secretion occurs via BK channels. *J Am Soc Nephrol* 17: 1275–82, 2006.

[104] Schneeberger EE, and Lynch RD. The tight junction: a multifunctional complex. *American Journal of Physiology* 286: C1213–28, 2004.

[105] Schroeder BC, Waldegger S, Fehr S, Bleich M, Warth R, Greger R, and Jentsch TJ. A constitutively open potassium channel formed by KCNQ1 and KCNE3. *Nature* 403: 196–9, 2000.

[106] Schultheis PJ, Clarke LL, Meneton P, Miller ML, Soleimani M, Gawenis LR, Riddle TM, Duffy JJ, Doetschman T, Wang T, Giebisch G, Aronson PS, Lorenz JN, and Shull GE. Renal and intestinal absorptive defects in mice lacking the NHE3 Na^+/H^+ exchanger. *Nature Genetics* 19: 282–5, 1998.

[106a] Schultz SG. Homocellular regulatory mechanisms in sodium-transporting epithelia: avoidance of extinction by "flush-through." *Am J Physiol* 241: F579–90, 1981.

[107] Schweinfest CW, Spyropoulos DD, Henderson KW, Kim JH, Chapman JM, Barone S, Worrell RT, Wang Z, and Soleimani M. slc26a3 (dra)-deficient mice display chloride-losing diarrhea, enhanced colonic proliferation, and distinct up-regulation of ion transporters in the colon. *Journal of Biological Chemistry* 281: 37962–71, 2006.

[108] Sellin J, and Duffey M. *Mechanisms of Intestinal Chloride Absorption*. New York: Raven Press, 1990.

[109] Sellin JH, and De Soignie R. Regulation of Na-Cl absorption in rabbit proximal colon in vitro. *American Journal of Physiology* 252: G45–51, 1987.

[110] Sellin JH, and DeSoignie R. Rabbit proximal colon: a distinct transport epithelium. *American Journal of Physiology* 246: G603–10, 1984.

[111] Suares NC, and Ford AC. Prevalence of, and risk factors for, chronic idiopathic constipation in the community: systematic review and meta-analysis. *American Journal of Gastroenterology* 106: 1582–91; quiz 1581, 1592, 2011.

[112] Tomson FL, Koutsouris A, Viswanathan VK, Turner JR, Savkovic SD, and Hecht G. Differing roles of protein kinase C-zeta in disruption of tight junction barrier by enteropathogenic and enterohemorrhagic Escherichia coli. *Gastroenterology* 127: 859–69, 2004.

[113] Vanner S, and Macnaughton WK. Submucosal secretomotor and vasodilator reflexes. *Neurogastroenterology and Motility* 16 Suppl 1: 39–43, 2004.

[114] Venkatasubramanian J, Rao MC, Sellin JH. Intestinal Electrolyte Absorption and Secretion. In: *Sleisenger and Fordtran's Gastrointestinal and Liver Disease* (9 ed.), edited by Feldman M, Friedman, LS, Brandt, LJ. : Saunders Elsevier, 2010, pp. 1675–94.

[115] Venkatasubramanian J, Sahi J, and Rao MC. Ion transport during growth and differentiation. *Annals of the New York Academy of Sciences* 915: 357–72, 2000.

[116] Venkatasubramanian J, Selvaraj N, Carlos M, Skaluba S, Rasenick MM, and Rao MC. Differences in Ca(2+) signaling underlie age-specific effects of secretagogues on colonic Cl(-) transport. *Am J Physiol Cell Physiol* 280: C646–58, 2001.

[117] Verkman AS. Aquaporins: translating bench research to human disease. *Journal of Experimental Biology* 212: 1707–15, 2009.

[118] Verkman AS. Drug discovery and epithelial physiology. *Curr Opin Nephrol Hypertens* 13: 563–8, 2004.

[119] Verkman AS. Knock-out models reveal new aquaporin functions. *Handbook of Experimental Pharmacology* 359–81, 2009.

[120] Walker NM, Simpson JE, Yen PF, Gill RK, Rigsby EV, Brazill JM, Dudeja PK, Schweinfest CW, and Clarke LL. Down-regulated in adenoma Cl/HCO3 exchanger couples with Na/H exchanger 3 for NaCl absorption in murine small intestine. *Gastroenterology* 135: 1645–53 e1643, 2008.

[121] Wang KS, Ma T, Filiz F, Verkman AS, and Bastidas JA. Colon water transport in transgenic mice lacking aquaporin-4 water channels. *American Journal of Physiology* 279: G463–70, 2000.

[122] Wapnir RA, and Teichberg S. Regulation mechanisms of intestinal secretion: implications in nutrient absorption. *Journal of Nutritional Biochemistry* 13: 190–9, 2002.

[123] Weihrauch D, Kanchanapoo J, Ao M, Prasad R, Piyachaturawat P, and Rao MC. Weanling, but not adult, rabbit colon absorbs bile acids: flux is linked to expression of putative bile acid transporters. *American Journal of Physiology* 290: G439–50, 2006.

[124] Weinman EJ, Cunningham R, and Shenolikar S. NHERF and regulation of the renal sodium-hydrogen exchanger NHE3. *Pflugers Archiv* 450: 137–44, 2005.

[125] Wood JD. Enteric nervous system: sensory physiology, diarrhea and constipation. *Current Opinion in Gastroenterology* 26: 102–8, 2010.

[126] Wright EM, and Loo DD. Coupling between Na+, sugar, and water transport across the intestine. *Annals of the New York Academy of Sciences* 915: 54–66, 2000.

[127] Wright EM, Loo DD, Hirayama BA, and Turk E. Surprising versatility of Na+-glucose cotransporters: SLC5. *Physiology (Bethesda, Md)* 19: 370–6, 2004.

[128] Xu H, Zhang B, Li J, Chen H, Tooley J, and Ghishan FK. Epidermal growth factor inhibits intestinal NHE8 expression via reducing its basal transcription. *Am J Physiol Cell Physiol* 299: C51–7, 2010.

[129] Xu H, Zhang B, Li J, Chen H, Wang C, and Ghishan FK. Transcriptional inhibition of intestinal NHE8 expression by glucocorticoids involves Pax5. *American Journal of Physiology* 299: G921–7, 2010.

[130] Yang B, Song Y, Zhao D, and Verkman AS. Phenotype analysis of aquaporin-8 null mice. *Am J Physiol Cell Physiol* 288: C1161–70, 2005.

[131] Ye D, Guo S, Al-Sadi R, and Ma TY. MicroRNA regulation of intestinal epithelial tight junction permeability. *Gastroenterology* 141: 1323–33, 2011.

[132] Zachos NC, Tse M, and Donowitz M. Molecular physiology of intestinal Na^+/H^+ exchange. *Annual Review of Physiology* 67: 411–43, 2005.

[133] Zeuthen T. Water-transporting proteins. *Journal of Membrane Biology* 234: 57–73, 2010.